HAMLET'S UNIVERSE

HAMLET'S UNIVERSE

Peter Usher

Aventine Press

Cover by RTP Signs & Graphics
Images from Art Explosion © 1995-2002

Published by Aventine Press
1023 4th Ave #204
San Diego CA, 92101
www.aventinepress.com

ISBN: 1-59330-444-7

Library of Congress Control Number: 2006938611
Library of Congress Cataloging-in-Publication Data
HAMLET'S UNIVERSE

Printed in the United States of America

Table Of Contents

PREFACE TO FIRST EDITION

The Sun-centered model of the planetary system advanced in 1543 by Nicholas Copernicus, marks the beginning of the age of the New Philosophy. Following Coffin (*Philosophy* 65), I take that term to mean the accumulation of facts about nature and the development of a critical and objective method of understanding phenomena of the physical Universe in the post-Copernican era. I restrict its meaning to the celestial sciences and use the term "New Astronomy" in the orthodox definition to mean the study of the forms of motion of celestial objects. "New Physics" describes theories on the nature of matter gleaned from terrestrial and celestial data. The term "Old" applies to these disciplines in the pre-Copernican era.

Chapter 1 introduces the ancient model of the Universe that placed the Earth at its center. According to this Earth-centered or geocentric scheme, crystalline spheres held the Sun, Moon and planets, and all rotated at different rates about the Earth. An outermost sphere held the stars and it rotated about the Earth as well. In the second century AD, the Greco-Roman astronomer, Claudius Ptolemy, refined this model, which remained essentially unchanged into the sixteenth century. In 1543, Copernicus advanced a radically new way to look at the planetary system by placing the Sun at the center and relegating the Earth to the rank of a planet. Only the Moon remained in orbit around the Earth. The outer bound of the Copernican model was a sphere of stars, as was that of another model introduced in the 1580s by the Danish astronomer Tycho Brahe. In 1576, Thomas Digges added a further dimension when he suggested that stars were scattered through infinite space.

Chapter 2 reviews, briefly, how some English poets of the early modern era interpreted the Sun, Moon, planets, and the Milky Way, and it examines the extent to which Shakespeare incorporated contemporary cosmic knowledge into his works. Chapter 3 describes the state of astronomical knowledge in England in the sixteenth century and examines the meaning of the term "infinite" in connection with the distribution of stars. Thomas Digges may have used his father's invention, the perspective glass, to observe stars and other celestial objects. Chapter 4 traces the early history of the science of optics and notes that Digges' father was a direct beneficiary of the work of Roger Bacon. Fundamental to the success of any worldview is the means by which inquirers interrogate their environment and reach conclusions. Chapter 5 looks into this subject, with special attention to fallacious thinking and faulty inference.

Shakespeare is fond of ambiguity and finds an ideal outlet in an equivocating prince, Amleth, whose escapades the medieval scholar, Saxo Grammaticus, recorded in his history of Denmark. Chapter 6 describes similarities between Amleth and Hamlet, notably, their metaphysical insights and alleged madness. Prince Hamlet's way of acquiring information differs from the deceitful practices of the Danish king, Claudius, and his henchman Polonius. Chapter 7 describes how Hamlet gets at the truth via a play-within-the-play. Chapter 8 elaborates on the mooted allegorical hypothesis and its personifications, and the relation between the script and the history of science and astronomy. The chapter moves on to the final phases of the plot.

The final four chapters explore consequences of the previous eight. Chapter 9 presents other apt identifications and Chapter 10 presents the case for Shakespeare's description of celestial phenomena that no one could have known unless aided by a telescopic device. Chapter 11 gives a novel interpretation of Hamlet's well-known remark concerning wind directions and sanity. Chapter 12 interprets Hamlet's love letter as a statement of the ascendancy of the New Philosophy, and the Afterword contains concluding remarks. Preparatory work is traceable through refereed papers and abstracts presented and published between 1996 and 2006 (see Works Consulted).

Throughout, I let "Canon" refer to the body of works by the Bard, William Shakespeare. Many believe that this name refers to William Shakspere (1564-1616), an actor who brought fame to Stratford-on-Avon, but others believe that the evidence for this identification is scant and that the name is a nom de plume for Edward de Vere (1550-1604), 17th Earl of Oxford.

I assume that the reader has a working familiarity with the Canon and with *Hamlet* in particular. Q1 of *Hamlet* refers to the so-called "bad" quarto, Q2 to the Second Quarto of about 1601, and F1 to the First

Folio edition of 1623. I use Edwards' *Hamlet* because it is a convenient amalgam of Q2 and F1. References to act, scene and line use Arabic numerals according to the convention "aa.ss.ll." Among several sources, I use the Dictionary of National Biography (*DNB*) for biographical details and I consult the Oxford English Dictionary (*OED*) for English usage during and before the sixteenth century. Context decides whether personal pronouns refer to both genders. Passages from Digges' *A Perfit Description* come from its reprinting by Johnson and Larkey. I capitalize object names and technical terms and regard "Cosmos," "Universe" and "World" as virtually synonymous. "World view" refers to any conception of the Universe that is necessarily part of a worldview. This work is multidisciplinary and I beg forgiveness for foraging in others' fields. Needless to say, all errors and omissions are entirely my own.

I am grateful to the American Astronomical Society and the Shakespeare Oxford Society for their professionalism and tolerance of interdisciplinarity. I thank Judd Arnold, Nancy Brown, Robert Chapman, Margaret Chester, Margaret Christian, Carl Croushore, Floyd Dorrity, Gordon Fleming, Gary Goldstein, Nina Green, Stephanie Hughes, Anna Jangren, Bruce Kendall, Alan Knight, Robert Lima, Steven May, Allan Mills, Gary Moorman, George Musser, Robert Naeye, Don Neidig, Peter Nockolds, Steve Sohmer, Julia Usher, Jan van der Meulen, Stewart Wignall, Daniel Wright and Linda Woodbridge, variously, for interest, help and encouragement. I thank anonymous reviewers who supported my work at crucial junctures. By far my greatest debt is to my wife. She gave me the strength to carry on and it is no exaggeration to say that, without her love and forbearance, this work would not have seen the light of day.

PREFACE TO SECOND EDITION

I take this opportunity to correct some minor errors.

CHAPTER 1: INTRODUCTION

It is much easier to propose rather speculative theories than it is
to get to a deeper understanding of even one simple elementary
problem.

Geoffrey Burbidge

Celestial phenomena play a central role in everyday existence, as
in the rising and setting of the Sun and its relation to the seasons. Sun
and Moon play a significant role in marking the passage of time and
knowledge of their cycles is important to survival. The ancient Greek
authors Hesiod (*fl.* 8th century BC) and Homer (*fl.* 6th century BC)
wrote of phenomena in the sky and their relation to events on Earth, as
did Virgil (70-19 BC) in *Georgics*, but by modern standards progress
toward understanding the physical heavens was extremely slow.

Apart from occasional impacts of meteorites, like the one that
caused a sensation when it fell on the Peloponnesus in 467 BC, early
studies of the sky were restricted merely to observations from Earth.
Even today, there are relatively few ventures in extraterrestrial space
that are subject to controlled experimentation as occur routinely in
ground-based sciences like agriculture and zoology. The intangibility of
celestial objects impeded the understanding of the physical heavens and
forced ancient scholars to rely on rational argument, which, in hindsight,
they were ill equipped to do. Argument that proceeds directly from
observation to explanation has a potential for error because a reliance on
reasoning to the exclusion of empirical verification of hypotheses can
lead to false conclusions. As a result, cause and effect became muddled
and magical thinking flourished.

This book addresses the revolutions in scientific thought that
occurred in the sixteenth century and presents evidence that suggests
that the modern age of systematic telescopic discovery began in that
century, decades earlier than generally believed. To pave the way, it is
helpful to outline the development and state of astronomical knowledge
up to that time.

Plato.

The fundamental question of the shape of the Earth was still uncertain at the time of Socrates (469-399 BC), but Plato (428-347 BC) settled the issue in favor of sphericity in conformity to what were then the two basic astronomical shapes, the circle and sphere. The earliest and most enduring of all cosmic concepts is the two-sphere Universe, in which stars lay on a perfectly spherical dome of sky and moved overhead on perfectly circular paths. These rudimentary observations, coupled with a belief in the sanctity of the heavens, led to the doctrine of the perfection of the circle and the sphere.

Plato lists astronomy as an essential part of education and mathematics and connects astronomy and religion by proclaiming that the study of the visible heavens is an antidote to atheism. *Timaeus* describes the creation of the Universe by the master Artificer whose beneficence, Plato believed, was essential to cosmic understanding. Outside the shell of stars lay the limitless space of the Empyrean, which was the true and ultimate paradise. Mortals were limited to observations in bounded space, with the rest off-limits to secular inquiry.

Plato had little time for empiricism because he believed that the evidence of the senses played a secondary role to the mind's power to envision an ideal World. The difficulty is that the vast majority of human minds were incapable of reconciling the perfections of their imagination with the reality of their experience. Most took celestial phenomena at face value, believing that what they saw represented what was real. Plato's Earth-centered, or geocentric, Universe was essentially poetical and intended more as a simulacrum for philosophical and spiritual guidance than as a picture of celestial reality, but one would think that a philosopher concerned with the spiritual heavens would have paid closer attention to celestial phenomena especially since most did not fit the imagined ideals of heavenly perfection.

Planets.

The term "Fixed Stars" describes stars that appear to move around the Earth but do not move relative to one another. The "Ancient Planets" are the seven objects, Sun, Moon, Mercury, Venus, Mars, Jupiter, and

Saturn, that turn about the Earth once a day but also wander relative to one another and to the Fixed Stars. Ancient astronomers called them "planets" from the Greek word for "wanderer," and they lumped Ancient Planets and Fixed Stars into one category, "star," because they believed they were all made of the same quintessential substance. Crystalline spheres held the Ancient Planets in place and an eighth sphere held the Fixed Stars. These spheres were transparent in order to allow us to see beyond them to the next outermost Wanderer and, thus, to the outermost sphere of stars, but one magical crystalline substance did not suffice because the starry Firmament had to be opaque to the brilliance of the Empyrean that lay beyond it.

The farther an object is, the slower it seems to move, prompting ancient Babylonians to arrange the Ancient Planets according to the time they take to complete a circuit of the sky:

(A)   Mercury, Venus, Sun, Mars, Jupiter, Saturn.

The fastest and nearest Wanderer is the Moon, which takes about 27 days to complete a circuit, and the slowest and farthest is Saturn, which takes nearly 30 years to complete a circuit. Mercury and Venus always lie close to the direction of the Sun and, on average, have the same circuit time of one year, but the Babylonians opted to put the Sun in the middle of sequence (A).

By the doctrine of First Cause, the *Primum Mobile*, or Prime Mover, impels the sphere of Fixed Stars to move, which, by magical connections, impels the sphere of the outermost Ancient Planet, Saturn, to move, but more slowly, as if there were slippage of the clutch that engages the invisible gears of the celestial machinery. Thus, as the sky rotates from east to west, the outermost of the Ancient Planets, Saturn, moves more slowly in the same direction, giving the appearance of drifting in the opposite direction, eastward, relative to the Fixed Stars. This process of apparent slippage repeats all the way down to the nearest Ancient Planet, the Moon, for which the accumulated slippage is greatest. This gives the impression that the Moon moves most rapidly eastward relative to the stars, although, in the supposed reality of the time, it has the least absolute motion.

The downward trend toward immobility ends at the corrupt Earth and its muddy vesture of decay. To observers on the ground, the show above is like theater-in-the-round because the patron thinks that he has the best seat in the house – at the dead center of the performance. From this special site, the clapper-clawed groundling views the cast of celestial players performing before a starry proscenium, but the choreography is baffling because, for no rhyme or reason, the ambits of the Ancient Planets deviate from heavenly perfection. No Wanderer moves at a steady pace and five of them temporarily reverse their easterly direction of travel relative to the Fixed Stars (see Figure 1).

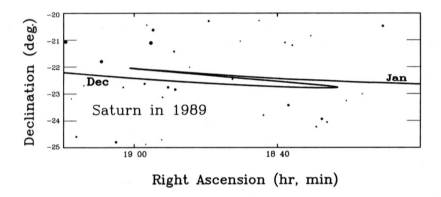

Figure 1: Retrograde motion of planet Saturn seen against distant stars.

Westward planetary motion relative to the stars goes by the name "retrograde motion" and afflicts only the star-like meanderers, Mercury, Venus, Mars, Jupiter and Saturn. Early philosophers were not concerned that three of them, Mars, Jupiter and Saturn, are always in the throes of retrograde motion when they are at their brightest and when, coincidentally, their positions in the sky are opposite to that of the Sun. Moreover, the other two star-like wanderers, Mercury and Venus, never stray by more than about 22° and 45° from the direction of the Sun. The two Ancient Planets that are immune to retrogradation also have easily discernable surfaces, signifying that they are either very close or very large, and, although the Moon's phases correlate with its position relative to the Sun and signify a geocentric orbit, the Sun has a relationship to the other five Ancient Planets that geocentric motion does not explain.

Eudoxus (408?-355? BC) accounted for planetary motions by devising a complex system of shells nested around a common center. He allowed each to rotate uniformly about an axis that connects to the next shell, which rotates uniformly about an axis that is connected and inclined to the next, and so on, for as many times as necessary to account for observations. Historians believe that the most sophisticated Eudoxian model was one devised by Plato's most famous pupil, Aristotle (384-322 BC), which had over fifty such spheres. Plato's emphasis on mathematics eventually caused mechanistic models like these to give way to numerical procedures, or algorithms, for calculating planetary positions, but retrograde motion proved a stubborn adversary and remained the premier mathematical and astronomical problem into the seventeenth century.

Widespread fear of heavenly influences and belief in celestial omens drove the need for planetary ephemerides (predictions of planetary positions in the sky) and inevitably, forced the symbolism of Plato's cosmic forms to confront the reality of visual perception. Plato had scant regard for empirical knowledge and he simply assigned to his students the task of working out the details. By the time a satisfactory solution emerged in the 1660s, Plato's homework assignment had occupied students for about two thousand years.

Aristotle.

Aristotle made positive contributions in all areas he studied but, by modern consensus, his influence in physics and astronomy was disastrous. He believed that data acquisition and systematic experimentation were infra dig and, although he considered empirical data, his efforts in the cosmic sciences were ill conceived and the product of a thinker who is uncertain about how to deal with extraterrestrial subject matter. He dutifully adhered to Plato's ideal of a spherical stationary Earth about which revolved the Firmament of stars and the seven Ancient Planets, but, like his mentor, he preferred to let others work out the details. His followers accepted his work uncritically even though he expressed doubts about the validity of some of it. In the centuries leading up to the seventeenth, weaker minds took shelter behind his authority and some were disinclined even to test his ideas lest they prove him wrong.

Thanks to Arab philosophers, Aristotle's works arrived piecemeal via Spain and Sicily to Western Europe. They underwent many translations along the way, during which adherents developed ideas of their own based on material available at the time. Various schools arose whose accumulated body of thought we now know as Aristotelianism. Thanks to Aristotle's fine reputation and impeccable academic lineage, the basic ideas behind the bounded geocentric World model went essentially unchanged and unchallenged into the sixteenth century.

Though he may have written other relevant material that is now lost, available information indicates that he did not give credence to the three fundamental cosmic possibilities that later became the basis of the New Astronomy. These are that the Earth rotates and moves through space and that the Universe is potentially infinite. He and his followers could not imagine that they were viewing the celestial drama from a rotating and revolving merry-mixer called Earth. They could not separate observed celestial motions into their basic components because they attributed all observed motions to the objects without realizing that they themselves were in motion. Early thinkers thought that what they saw was reality and made no allowance for the fact that they were located at the center of their own perception. In modern times, the pitfall of self-centeredness has given rise to the Principle of Location, which warns that it is unlikely that any observer has a special location in the Universe.

Aristotelians sought merely to "save appearances" in the sky by contriving mechanisms that could account for what they saw. The somewhat misleading expression "to save appearances" arose from a clumsy rephrasing of Aristotle's writing by Sosigenes (*fl.* 2nd century AD), but the term means essentially, to represent or explain phenomena or, as John Milton (1608-1674) writes in *Paradise Lost*, to "model" them.

Aristotle's cosmology is physical as well as astronomical. In accordance with hierarchical cosmic structure, Aristotle imagined that four elements, Earth, Water, Air and Fire accounted for all material. He supposed that each element predominated in concentric regions that were ordered with some overlap from the Earth outward into sublunary space and that a quintessential fifth element, Ether, holds sway in the superlunary realm of celestial perfection. Aristotle made an exception to this rule by ascribing starlight to the interaction of Fire and Air, which friction ignited as the Fixed Stars and Ancient Planets moved round the Earth.

Leucippus (fifth cent. BC) wrote *Great World System* sometime around 435 BC and, with his better-known student Democritus (*c.*460-*c.*370 BC), conceived of a Cosmos comprised of different sorts and sizes of atoms that arrange themselves naturally into the Universe we observe. Aristotle tried to discredit these ideas because they contradicted his theory and did not fit the orderliness of his hierarchical scheme.

Parallax.

The father of astronomy, Pythagoras (*c.*582-*c.*507 BC), and his followers Heraclitus (*c.*535-*c.*475 BC) and Ecphantus (fifth cent. BC) believed in a rotating Earth and, according to Cicero (106-43 BC), so did Hicetas of Syracuse (fifth cent. BC). Philolaus (fifth cent. BC) believed that the Earth revolved about the central fire and rotated at such a rate as always to keep the fire from view.

Aristotle recognized the possibility that the Earth might move, but argued against it on several grounds. If the Earth rotated or revolved and he were to look at two stars lying in a particular direction, he would be sometimes nearer and sometimes further away from them so that they would appear sometimes farther apart and sometimes closer together. The difference between these two angles is the parallax angle, which is larger and easier to detect if the Earth were to revolve than if it merely rotated because the Earth's orbit is necessarily larger than the Earth itself. He saw no such effect either daily or annually, implying either that the stars and Ancient Planets were extremely distant or that the Earth did not move. Aristotle argued that, if the stars were extremely distant, there had to be a large volume of space between them and Saturn's orbit. He believed that this unused space would contradict the doctrine of Final Cause, which held that creation was purposeful and that everything in the Universe served a function. This doctrine held that "place" was a location where something should reside, so Aristotle argued that a beneficent Creator would not construct a Universe with space that was full of emptiness and, therefore, that the Earth was stationary.

The steadfastness of Aristotle's convictions rested upon the confidence that he had in developing his positions and reaching his conclusions. His collected works on logic are called *Organon,* of which *Categories* is a part, but, in the sciences at least, his logic is deficient and the categories to which he assigned the existents of the Universe are wanting. For example, in the matter of categories alone, he fails to address anomalies in the nature and motion of Ancient Planets and lumps them all into a single, undifferentiated group.

The Celestial Sphere.

The Celestial Sphere is a representation of the sky upon which observers place positions of celestial objects. To believers in bounded geocentricism, it represented the rigid, opaque, spherical encasement of the material world, but nowadays we imagine that it is a sphere of arbitrary size, centered on the observer, whose "surface" shows the directions of stars and planets regardless of their actual distance.

Celestial coordinates identify those directions and mimic those that geographers use. Geographers measure Latitude north and south of the Equator, and Longitude east or west of the Greenwich meridian. To get analogous celestial coordinates, we imagine the Earth's Equator projected outward to form the Celestial Equator from which astronomers measure the celestial coordinate, Declination. Similarly, the Earth's axis projects outward to meet the Celestial Sphere at the North and South Celestial Poles. Both celestial and terrestrial Equators divide their respective spheres into Northern and Southern Hemispheres.

During the year, the Sun appears to move progressively into more easterly constellations. The Sun's path defines the Ecliptic, so-called because eclipses of the Sun must occur along it. If it were possible to locate ourselves on the Sun and look at the Earth, we would see that it follows the same Ecliptic path against the background stars over the course of a year. The Moon and planets appear projected against the Zodiacal constellations because their orbital planes are only slightly inclined to one another and to the plane of the Earth's orbit, so that, when seen from Earth, they all appear near the plane of the Sun's path, i.e., close to the Ecliptic. The Celestial Equator and Ecliptic define planes of indefinite extent that cut the Universe into two parts. The planes do not coincide but incline to one another by an angle, 23½°, called the Obliquity of the Ecliptic, which, for dynamical reasons, remains effectively the same through time. [A right angle = 90°, 1° = 60 minutes of arc, 1 minute of arc = 60 seconds of arc.]

The four astronomical seasons are Spring, Summer, Autumn and Winter, which, by convention, coincide with the four meteorological seasons experienced in the Northern Hemisphere. At the start of Spring, March 21, the Sun is on the Celestial Equator and shines vertically down on the Earth's Equator. At the start of Summer, June 21, the Sun is at its maximum, 23½°, north of the Celestial Equator and shines directly down on the Earth's Tropic of Cancer. September 23 marks the start of Autumn (Fall) when the Sun shines down on the Earth's Equator again. On December 21, at the start of Winter, it shines directly down on the Tropic of Capricorn located at Latitude 23½° south. The positions of the Sun when it shines directly down on the Earth's Equator are the Equinoxes, so-called because on those days the Sun is above and below the horizon for equal amounts of time. The Solstices are so-called because then the Sun's path along the Ecliptic is instantaneously stationary in the sense that when the Sun reaches its extreme position north or south of the Celestial Equator it pauses before reversing direction. At the time of the Winter Solstice, nights are longest in the Northern Hemisphere and shortest in the Southern, and six months later, at the Summer Solstice, the opposite occurs. The celestial coordinate, Right Ascension, is analogous to Longitude on Earth and both need a zero-point from which to measure angle. By convention, the Greenwich meridian serves as the zero-point for Longitude and Right Ascension is measured from the point on the Equator where the Sun crosses the Equator headed north.

The Sun passes through twelve constellations in the course of its annual circuit along the Ecliptic, eleven of which bear the name of some sort of creature. Perhaps in recognition of the Sun's role in sustaining life, this band of constellations goes by the name "Zodiac," from the Greek word for animal. The only non-zoological name is Libra, the Scales, which is aptly named since, at the Autumnal Equinox, the Sun is above and below the horizon for equal amounts of time, as if balanced on a scale.

Precession.

In 134 BC, while determining the celestial coordinates of stars, Hipparchus (*fl.* 130 BC) discovered a phenomenon known as the Precession of the Equinoxes. This is a gyroscopic effect like the wobble of a toy top spinning at an angle to the vertical. The rotating Earth has an equatorial bulge that does not lie in the plane of its revolution and, as a result, the Sun exerts a differential gravitational torque that tries to align the Equator with the plane of the Ecliptic. However, the Earth maintains its obliquity to the Ecliptic and precesses instead, resulting in the Celestial Equator sliding along the Ecliptic and the Celestial Poles executing circles around the vertical to the Ecliptic. One precessional cycle lasts about 26,000 years, which corresponds to a rate of one

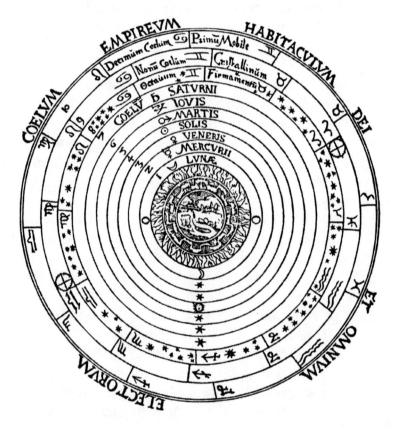

Figure 2: The bounded geocentric model of the Universe according to Claudius Ptolemy. (From Apian *Cosmographia*.)

Zodiacal constellation about every 2,200 years, or about 50 seconds of arc per year. When Hipparchus discovered Precession about 2,200 years ago, the Sun was in Aries when it crossed the Equator headed north. This intersection point retains the technical name, "First Point in Aries," even though the actual intersection point moves continuously through all twelve Zodiacal constellations. In the twenty-first century, the First Point in Aries has moved out of Aries and into Pisces, and is about to enter Aquarius.

Ptolemy.

Plato's emphasis on mathematics prompted quantification of the planetary model and led to models of increasing mathematical complexity as consumers of celestial forecasts demanded higher accuracy. Prior to the onset of the Dark Ages, Claudius Ptolemy (90-168 AD) was the last astronomer of note to make significant improvements in the art of forecasting celestial events. His thirteen-volume *Almagest* is a comprehensive account of astronomical data and theory and owes much to the work of his predecessors, notably Hipparchus. Figure 2 is a cartoon by Peter Apian (1495-1552) from the early sixteenth-century that shows the basic form of the Aristotelian-Ptolemaic Universe.

In order to account in detail for the motions of the Wanderers, Ptolemy allowed each to move along a circular Epicycle whose center moved along a circular Deferent (see Mitton 137-8). He offset the Earth from the center of the Deferent by the amount called the Eccentric, and created an Equant point that he displaced by an exactly equal amount on exactly the opposite side of the center of the Deferent. The Equant became the new center of the uniform angular motion of the center of the Epicycle. Ptolemy applied these rules ad hoc and adjusted them so that, when compounded, the result mimicked the actual motion of a Planet. Because there were no rules to indicate how to select and manipulate the geometry, the Ptolemaic model is more an algorithm geared to saving phenomena through calculation than a system based on a self-consistent pattern of explanation that could serve as a basis for physical understanding.

In an attempt to "save" the phenomenon of Precession, Ptolemy added a ninth sphere to the eight that bore the Ancient Planets and the stars. He connected the ninth sphere to the eighth, and compelled it somehow to rotate once every precessional cycle. Later astronomers added a tenth sphere to account for a supposed variation in Precession called Trepidation, and others added even more.

The accuracy of Ptolemy's model declined over time. In Toledo in 1080, scholars recalculated planetary ephemerides and repeated the process in 1252 under the supervision of King Alfonso X (1221-1284). The so-called Alfonsine Tables remained competitive into the seventeenth century but, overall, they predicted planetary positions to an accuracy of only about 1°, which is twice the angular size of the Moon.

Size of the Ptolemaic Universe.

The first step in achieving a three-dimensional view of extraterrestrial space is to determine the size of the Earth. One of the first to do so was Eratosthenes (*c*.276-*c*.195 BC), who headed the library at Alexandria, Egypt. He noticed that, at midday on the first day of summer, a vertical gnomon cast a shadow of 7¼° whereas at the same time, 7¼° further south at Syene, a vertical gnomon cast no shadow at all. This signified that Syene lay on the Tropic of Cancer. Alexandria is almost directly north of Syene, so Eratosthenes knew that the arc of the Earth's surface of length D from Alexandria to Syene was in the same proportion to the circumference of the Earth, C, as 7¼° is to 360°. Thus, $C/D = 50$, so to get C, Eratosthenes needed the distance D between Alexandria and Syene. He used the known fact that the average time for a camel to travel between the two places is 50 days and that a camel travels at an average rate of about 100 stades per day. Thus, in 50 days it would travel 5,000 stades. This gives D in units of a camel-day and C is, therefore, 50 times this value, or about 250,000 stades. To get the radius of the Earth R, he divided C by 2π, and found $R = 40,000$ stades, or thereabouts. The actual value of a stade in modern units is uncertain, but if it is about a tenth of a mile (176 yards, or roughly the length of a "stadium"), then the radius of the Earth is about 4,000 modern miles, close to the true value. Alternatively, if we use Ptolemy's value for a stade of about 155 yards, then the Earth's radius becomes slightly smaller.

To measure extraterrestrial sizes and distances in units familiar to us on Earth, early astronomers used the known value of the Earth's radius as a yardstick. Historically speaking, this is equivalent to projecting the motor ability of an average camel into cosmic space. Aristarchus of Samos (c.310-250 BC) began the process. He used ingenious geometry to get the Earth-Sun distance, but the data he used were less than satisfactory. For example, he believed that the direction of the Sun was 87° away from the Quarter Moon even though, in fact, this value is so close to 90° that the true value is difficult to measure. The angular diameters of the Moon and Sun are virtually the same as evident at the time of solar eclipses and, even though Aristarchus knew that each had an angular size of ½°, for obscure reasons he adopted a value of 2°. These lapses arose from a lack of "scientific" methodology and a corresponding inability to distinguish mathematical hypothesis from empirical fact. As a result, Aristarchus found that the Sun was only 18 to 20 times further from Earth than the Moon and since Sun and Moon have about the same angular size, the Sun had to be 18 to 20 times larger than the Moon, which made it only about 7 times larger than the Earth. In modern fact, the Sun is about 400 times farther from Earth than the Moon and 400 times larger. Astonishingly, these early underestimates survived for the better part of two millennia, and were current in the sixteenth century.

By clever geometry, Hipparchus determined the sizes and distances of the Moon and Sun in units of the size of the Earth, enabling him to assign terrestrial units to celestial distances. Claudius Ptolemy, using a slightly different approach, somehow ended up confirming the earlier results, leading some to question his methodology and data, if not his intellectual honesty. The net outcome was that Ptolemy and Arabic astronomers of the ninth and tenth centuries put the Earth-Sun distance at about 1,210 Earth radii (E.r.), and the sphere of stars at about 20,000 E.r. For the unit distance of 1 E.r. they adopted a value of about 3,250 modern miles, so that their model of the Universe had a radius of about 65 million modern miles. In other words, they thought that all of creation was jammed into a sphere of radius about seven-tenths of the actual value of the Earth-Sun distance.

Aquinas.

Thomas Aquinas (1225-1274) became convinced of the superiority of Aristotelian philosophy and set out to unify it and Christian dogma in the belief that knowledge and faith should both serve the ends of truth. He studied in Paris under Albertus Magnus (1193-1280) and, after time spent in Cologne, he returned to Paris as a Professor of Theology. Aquinas accepted Aristotle's argument for a Prime Mover because he believed that motion had to arise from prior motion and that it was nonsensical to attribute motion to prior motion indefinitely. The same sort of argument pertained to physical existents whose form could derive from earlier existents through material transformation, but that process had to start somewhere as well, and there had to be a Creator of matter in the first place.

Aquinas' argued his chief work, *Summa Theologiae,* so finely that the Christian Church accepted its precepts and, as a result, bounded geocentricism became enshrined in doctrine and ensconced in the monasteries of learning. This thirteenth-century synthesis was a great achievement of medieval philosophy and became known as Scholasticism. In building a natural theology based largely on Aristotelian thought, the so-called schoolmen sowed the seeds of conflict because, with the advance of the New Philosophy, it became clear that scientific theory is provisional whereas theological truth is absolute.

The Aristotelian model had theological appeal because, among other reasons, there was an easily understood division between physical and metaphysical space. Mystical beliefs owed much to Plotinus (205-270) and later neo-Platonists who believed that a hierarchy of angels made their homes in the divine spheres of the planets that ascended successively outward to the home of the Prime Mover. The notion that Heaven is "up" and Hell is "down" was current long before the Scholastic synthesis. Boethius (480-525) called the substance of God the sphere of the Fixed Stars. The sphere of the Earth enabled Anaxagoras (*c*.500-*c*.428 BC) to comfort a man who was dying in a foreign land by telling him that the descent to Hell is the same from every place. Later, in 1588, during a debate on the Inferno of Dante Alighieri (1265-1321), Galileo Galilei (1564-1642) made fun of Dante's geocentricism by explaining that God placed the Earth at the center of the Universe to have it as far as possible from the sight of the blessed residents of Heaven lest its grossness offend them.

As Thomist doctrine incorporated Aristotelian thought, so it imported also the celestial calculus of Ptolemy, which remained the standard cosmological model into the seventeenth century. Nevertheless, a few early thinkers had speculated that the Earth did not lie immobile at the center of the Universe. These included Heraclides Ponticus (c.388-c.315 BC), Aristarchus of Samos, Apollonius of Perga (*c.*262-190 BC), Hipparchus, Seleucus the Babylonian (2nd cent. 150 BC), Seneca (4 BC-46 AD), Aryabhatta (*fl.* 5th cent), Brahmagupta (*fl.* 7th cent), Bernardus Silvestris (*fl.* 1147), Nicole Oresme (1320-1382) and Nicholas of Cusa (1401-1464). By the sixteenth century, discontent with Aristotelian philosophy was on the rise and, in 1536, Petrus Ramus (1515-1572) suggested that everything that Aristotle said was wrong and that knowledge gleaned from a combination of empirical evidence and rational argument should replace it. The culture of the Renaissance was conducive to change and, in 1543, in cosmology, it was the lot of Nicholas Copernicus (1473-1543) to effect it.

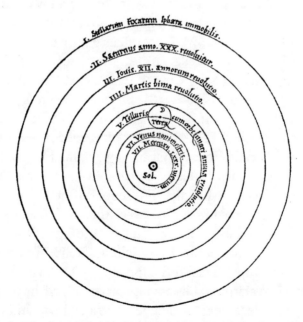

Figure 3: The bounded heliocentricism model of the Universe according to Nicholas Copernicus. (From *De Revolutionibus.*)

Copernicus.

Copernicus was born in Torun on the Vistula in Poland. After attending the University of Cracow, he studied law and medicine in Italy. He devoted effort to astronomy as well, coming under the influence of a Professor of Mathematics and Astronomy at the University of Bologna, Domenico Maria da Novara (1454-1504), who was an ardent neo-Platonist and defector from the camp of Ptolemaic astronomy. Copernicus returned to Poland where he served as a church administrator and where his cosmological apostasy began in earnest. He subscribed to the Pythagorean view that the Cosmos adhered to rules that were capable of mathematical and aesthetically pleasing expression. So guided, he developed a Sun-centered, or heliocentric, planetary model that was unencumbered by the ad hoc mechanisms that burdened geocentricism (see Figure 3). He left the Moon to orbit the Earth, but argued that the Earth rotates daily on its axis and revolves about the Sun. The resulting Solar System has the Earth in the company of the remaining five Ancient Planets, all six of which we now call, simply, "planets." Earth is no longer special but is a planet in its own right and allowed, therefore, to have planet-like properties, including orbital motion.

In *Commentariolus* of 1529, Copernicus gave a preliminary account of orbital revolution, which, seven years later, came to the attention of a prominent official of the Church who asked Copernicus to make his theory fully known to the world. In 1539, Georg Joachim (1514-1576), who called himself Rheticus, went to Frauenberg to study under Copernicus. He risked entering a diocese that had just issued a proclamation on heretics, but Copernicus welcomed him and what was to have been a short visit lasted two years. In 1541, Rheticus returned to Wittenberg and established the first school of heliocentric planetary astronomy. This occurred thirty-nine years after its university's founding and twenty-four years after Martin Luther (1483-1546) had nailed his ninety-five theses to the door of the Schlosskirche. Andreas Osiander (1498-1552) saw Copernicus' full exposition, *De Revolutionibus*, into print but became so alarmed at its novelty that he added the words *Orbium Celestium* to the title to suggest that geocentricism was still the order of the day. As a further precaution, he inserted an unsigned prefatory note explaining that the work was algorithmic and not to be taken literally.

De Revolutionibus appeared in 1543, just before Copernicus' death, prompting the observation that, with felicitous timing, Copernicus published as he perished (Rosen "Copernicus").

In order from the Sun outward, the Copernican planets are:

(B)   Mercury, Venus, Earth, Mars, Jupiter, Saturn.

Sequence (B) differs from the old sequence (A) in two ways: Moon is absent because it retains its geocentric orbital property and now belongs to a brand new category of "planetary satellites," and Earth replaces Sun as the entry between Venus and Mars, which is tantamount to a transformation of center and thus to a re-definition of orbital geometry. Copernicus also retained the old concept that the circle was the basic orbital shape.

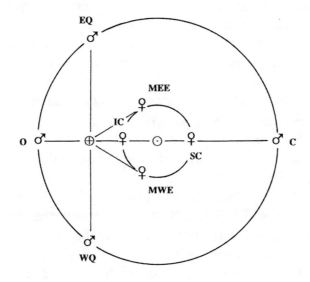

EARTH ⊕   SUN ☉   MARS ♂   VENUS ♀

SUPERIOR PLANETS:		INFERIOR PLANETS:	
O	Opposition,	IC	Inferior Conjunction,
C	Conjunction,	SC	Superior Conjunction,
EQ	Eastern Quadrature,	MEE	Maximum Eastern Elongation,
WQ	Western Quadrature.	MWE	Maximum Western Elongation.

Figure 4: Planetary alignments.

To accommodate observations from Earth, its position in (B) divides the other planets into two categories: Mars, Jupiter and Saturn are Superior Planets because their near-circular orbits lie outside the Earth's, whereas the Inferior Planets, Mercury and Venus, have orbits interior to the Earth's (see Figure 4). A moving Earth carries observers with it, so Copernicus must distinguish between appearance and reality. He writes, "why not admit that the appearance ... belongs to the heavens but the reality belongs to the Earth?" He explains that motions do not "belong" entirely to objects in the sky, but rather that they "borrow" some of the apparent motion from the movement of the Earth. By giving the Earth rotational and orbital motion, Copernicus lets it bear some responsibility for appearances in the sky, such as the daily rising and setting of the stars and the Sun's annual path through the Zodiac.

The most impressive Copernican achievement is his elegant solution to the problem of planetary retrogradation, which he explains as an appearance resulting from the orbital motion of a planet relative to an observer on the orbiting Earth. For example, an observer sees a Superior Planet, like Mars, moving retrograde from east to west because Earth moves on an inside orbital track and has greater speed. Only when the Earth is either well behind or well ahead, does that planet appear to run in the same direct sense, i.e., from west to east relative to the stars. In effect, the five geocentric epicycles required to "save" retrogradation in the Ptolemaic scheme fall victim to the single proposition that the Earth orbits the Sun.

Copernican theory also explains a property of the Superior Planets, best seen for Mars and Jupiter, that they are brightest around the time of Opposition and progressively less bright as they approach the same direction as the Sun. Moreover, the seeming affinity of Mercury and Venus for the Sun no longer requires the invention of a special category because all five unresolved Ancient Planets have heliocentric orbits. The Copernican solution has aesthetic appeal and is preferable to the Ptolemaic constructions because its economy of suppositions accords with the precepts of William of Occam (c.1280-1349) whose instrument of logic, "Occam's razor," states that, when formulating theory, hypotheses are not to be multiplied without necessity.

At Wittenberg, Erasmus Reinhold (1511-1553) calculated ephemerides based on heliocentricism and a Prussian nobleman paid for their publication, which became known as the Prutenic Tables. In England, John Dee (1527-1608) stated in his preface to *Ephemeris Anni 1557* of John Field (1525?-1587) that he had persuaded Field to revise the Prutenic Tables and calculate ephemerides suited for use in England. In the same year, Robert Recorde (1510-1558) produced *Castle of Knowledge* that hinted at the superiority of the heliocentric model.

Copernicus and Recorde noted precedents set by ancient believers in a moving Earth and saw Copernicus more as a revivalist rather than a revolutionary. The gadfly philosopher and priest, Giordano Bruno (1548-1600), also believed that the Copernican treatise was a restatement of Pythagorean truths, as did Galileo and the German mathematician Johannes Kepler (1571-1630). Despite the oftentimes whiggish interpretation of what modern historians loosely term the "Copernican Revolution," it cannot be said that heliocentricism was universally the model of choice in the sixteenth century, particularly since its accuracy was not much better than that of its geocentric competitor.

Fire.

Democritus, Plato, and later philosophers knew that the Moon borrows its light from the Sun and does not shine by Fire like the other stars. The old category of Ancient Planet is deficient right from the start because the Moon is unlike the other members and warrants a sub-category all its own. Copernican theory challenged the doctrine of Fire again by making Earth a Solar System planet because then, both Earth and Moon move yet they have hemispheres that are sunlit and unlit, and neither is afire. In addition, Copernicus put both Sun and stars to rest, which implies that they too, do not shine by Fire. The only candidates left that could possibly kindle Fire are Mercury, Venus, Mars, Jupiter and Saturn.

Although Copernicus did not completely extinguish Fire, the proliferation of exceptions does not bode well for its survival. Occam's razor and the new World view threatened it and, thereby, the foundation of the Old Physics. The radial hierarchy of the four material elements came under fire and, in particular, Earth was no longer a special substance whose primary property was a propensity to sink to the center of the World.

Immensity.

To map the Solar System, Copernicus needed two properties: relative scales for planetary orbits and an absolute scale of distance. He achieved the first by measuring angles and the intervals of time between various configurations of the Sun and planets, which gave him planetary orbits in units of (i.e., relative to) the Earth-Sun distance. His distance ratios are close to the modern ones, but his distance scale still relied on the ancient value for the Sun-Earth separation, which, as noted, is about twenty times too small.

Copernicus was well aware that no one had detected a stellar parallax, which, as we have seen, means either that the stars are very far away or that the Earth is at rest. He preferred to accept the Earth's motility than tolerate the hodge-podge of the Ptolemaic model, which meant that he had to accept the enormity of the Universe. He was not the first to encounter the problem. In 216 BC in *Sand-Reckoner*, Archimedes (*c.*287-250 BC) quotes the heliocentricist Aristarchus as saying that the sphere of the fixed stars centered on the Sun is so great that the orbit of the Earth "bears such a proportion to the distance of the fixed stars as the center of the sphere bears to its surface." Similarly, Copernicus declared that the distance of the stars is so great that, by comparison, the distance of the Earth from the Sun "is imperceptible." For Copernicus, the world within the vault of the stars is an *immensum* so that, by comparison, the Earth is "as a point." Although Copernicus realized that these very distant stars need not lie on a bounding sphere but could be scattered through space, he left "the philosophers of nature" to decide the matter and, like Aristarchus, opted for a Universe with a bound so large that he could not determine it.

With only his eye to guide him, Copernicus had no hope of establishing just how large the *immensum* really was. To gain perspective, consider that stars are so distant that heliocentric parallaxes of the nearest ones (that use the Earth's orbital radius as a baseline) are, at most, a fraction of a second of arc. Only after the invention of the telescope could angles this small be measured. In 1838, Friedrich Wilhelm Bessel (1784-1846) found a parallax of 0.29 seconds of arc for the star 61 Cygni and, the next year, Thomas Henderson (1798-1844) at the Cape of Good Hope found that α Centauri had a parallax of 0.75 seconds of arc, making it virtually the closest star to Earth and a mere 25,000,000,000,000 miles away. At that distance, light traveling at 186,000 miles per second takes a little longer than four years to reach the Earth.

Before these tiny parallax angles became measurable, James Bradley (1693-1762) had to discover the much larger effect of the aberration of starlight, which results from the finite speeds of light and of the Earth in its orbit. When we look at a star from the moving Earth, we see that it lies slightly closer to the point on the Celestial Sphere to which the Earth is moving than if the Earth were at rest. The same sort of phenomenon occurs when walking in rain, which, if it falls vertically, requires walkers to keep their feet dry by pointing the umbrella slightly forward in the direction of motion. In 1728-9, Bradley found that, to keep starlight shining down the barrel of his telescope, he had to point it forward by as much as 20 seconds of arc, a value nearly thirty times greater than the largest known stellar parallax. Discovery of aberration verified the Earth's revolution directly and came almost exactly two centuries after Copernicus first proposed the idea in *Commentariolus*. Experimental verification of the other Copernican proposition, that the Earth rotated, occurred in 1851 when Jean Foucault (1819-1868) constructed a freely swinging pendulum.

Tycho Brahe.

Tycho Brahe (1546-1601) was the survivor of twins born in Denmark on December 14, 1546. While still very young, his uncle and aunt abducted him and raised him as their own child so that, in effect, they were also his father and mother. (Historians commonly address Brahe by his Christian name, pronounced "Tee-ko.") In 1562, Tycho enrolled at the university in Leipzig where he engaged secretly in the study of astronomy. Crude measurements of the positions of planets soon convinced the teenager that neither Ptolemaic nor Copernican ephemerides were satisfactory.

In 1566, Tycho attended the University of Wittenberg where he spent a few months before fleeing an epidemic. He went to Rostock where he entered into a dispute with a third cousin. At issue, some think, was Tycho's analysis of a lunar eclipse, which, Tycho claimed, foretold the death of the Ottoman sultan, Suleiman the Magnificent (1494-1566), who, as luck would have it, had died six weeks before. During the fight, Tycho suffered the loss of his nose.

In 1572, a New Star appeared in Cassiopeia. Tycho first saw it on November 11, five days after Professor Wolfgang Schuler (d.1575) made the first sighting from Wittenberg. Tycho studied the star-like apparition, and showed that it lay farther away than the Moon. In 1573, he published his findings in *De Nova Stella* in which he posited that the New Star was far away like the Fixed Stars. This contradicted Aristotelian physics, which rested on the belief of heavenly immutability and perfection. Since the New Star underwent a sudden brightening followed by a steady decline, it followed that, in order for the sky to remain perfect, it would have to pass through a continuum of states resulting in a semantically and theologically challenging plurality of perfections.

Among historical events, Tycho's Supernova (as it is called, or SN 1572 in standard shorthand), and Kepler's Supernova, SN 1604, are the best documented of any up to that time for they occurred when interest in understanding the heavens was on the rise in Europe. Tycho's work so impressed the Danish monarch, King Frederick II (1534-1588) that, in 1576, he ceded to him the small Danish island of Ven (Hven) in order that he might build an observatory. Tycho completed his house in 1581 and named it Uraniborg, the Castle of the Heavens. Excavation of sites for large instruments began soon thereafter. Another building, Stjerneborg, the Castle of the Stars, was remarkable for its observation posts situated below ground level. While Tycho was establishing his observatory, the king was building Kronborg Castle in Helsingør a short distance north-north-west of Ven.

Tycho observed the Comet of 1577 and reported his results in 1588 in a limited first edition of *De Mundi aetherei recentioribus Phaenomenis Liber Secundus*, which he circulated at about the time as he started work on an introduction to the "New Astronomy," *Astronomiae Instauratae Progymnasmata*. In *Liber Secundus*, which was the Second Book of a planned trilogy, he showed that the comet lay beyond the sphere of the Moon, a result at odds with the popular notion that comets were atmospheric phenomena. The comet appeared to move with impunity among the planetary spheres of the Old Astronomy, prompting Kepler to write that Tycho destroyed the reality of the crystalline spheres.

Throughout his life, Tycho's positional measurements were limited by the visual acuity of his eye, which was, at best, about ½ to 1 arc minute. Various factors degraded accuracy further, but Tycho was able to measure star positions routinely to accuracy of about 2 minutes of arc. For a select group of nine stars, improved techniques led to accuracy twice as good, about 15 to 30 arc seconds, which was still too poor to detect the aberration starlight. Nonetheless, he was able to convince the world that the phenomenon of Trepidation, which had bedeviled astronomy since the start of the tenth century, was attributable to errors of observation.

As a first step in building a new model of the Universe, Tycho had to come to grips with the fact that he could not detect stellar parallax with the instruments he had available. On the Copernican hypothesis of a revolving Earth with an orbital radius equal to the Copernican value, 1,142 E.r., the lack of observed parallax led him to conclude that stars were more than 700 times further than the average distance of Saturn, which he put at 10,550 E.r. Since the volume of a sphere increases as the cube of its diameter, the volume of space surrounding the Solar System would have to exceed that of Tycho's model by a factor of 700 cubed, or over 300 million times. As a devout Aristotelian, Tycho considered that this was a ridiculous waste of space. In addition, he made the standard assumption that the appearance of a star in the sky was a measure of its true size and found that, at such great distances, the physical sizes of stars were hundreds to thousands of times the size of the Sun. He thought that this was absurd as well. Rather than entertain the possibility of what seemed to him stupid ideas, he opted for a stationary Earth, explaining that he wished to avoid the "physical absurdity" of the Copernican *immensum* by reducing everything to what he deemed was the Earth's stability.

If Tycho intended his motto *Non Haberi Sed Esse* ("not to seem but to be") to apply to the physical world, he did not follow it very well because, when it came to extracting a picture of reality from raw data, he was, like Aristotle, methodologically purblind. The irony is that, by failing to detect stellar parallax, Tycho had actually rendered a Copernican *immensum* plausible by placing a lower limit on its size that was enormous by contemporary standards. Lending credence to the *immensum* was not exactly Tycho's goal, however, because he had already decided on its antithesis, a *minutum*.

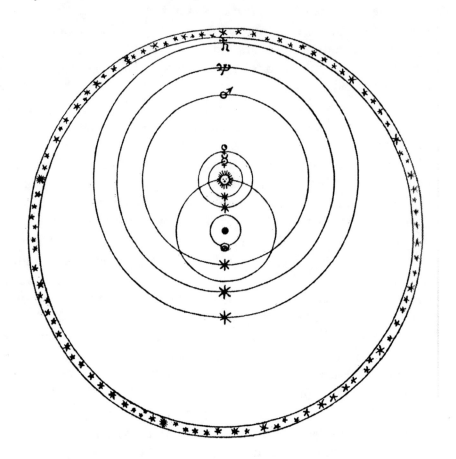

Figure 5: The bounded geo-heliocentric model
of the Universe according to Tycho Brahe.
(From *De Mundi Aetherei Recentioribus Phaenomenis*.)

At the same time, he rejected certain features of the standard Ptolemaic algorithm because he believed that, at Opposition, Mars was closer to Earth than the Sun. Since he regarded the Ptolemaic system as essentially correct and wished merely to dispense with its dysfunctional elements, he proposed a hybrid geo-heliocentric model of the Universe whose origin dates to lectures he delivered in 1574. He put the two resolved objects, Sun and Moon, into geocentric orbits and allowed the remaining five star-like Wanderers to orbit the Sun. With Earth and Sun both as centers of motion, his model is part geocentric and part heliocentric (see Figure 5), warranting the label "hybrid."

His choice for the nature and distribution of stars rested on the standard assumption that the apparent angular sizes of stars are measures of their actual size, so he put stars a mere 14,000 E.r. away, which meant that their physical diameters fell in a range of about 2/3 to 4 E.r. Like everyone else at the time, Tycho believed that the Sun was about 5½ times the size of the Earth, making stars smaller but comparable in size to the Sun, which he believed was reasonable. Tycho set the apogee of the outermost planet Saturn, at 12,300 E.r. and layered the rest of the Ancient Planets in a plenum in such a way as to optimize their concentration, the idea being, in accordance with Aristotelian teleology, not to "waste" space. Other geocentric models placed stars at distances of 19,000 to 20,100 E.r., making Tycho's model about three times less voluminous than Ptolemy's was.

It occurred to Tycho that the different apparent brightnesses of the stars could result from their being at different distances. He had found that bright stars were about 6 times the angular size of faint ones, so on the assumption that all stars had the same physical size, he thought that the faintest might lie, maybe, 6 times further than the brightest. He rejected such a large factor but made a token gesture by putting stars in a shell, which he guessed was about 1,000 E.r. thick. The Fixed Stars had to keep the same positions relative to one another as the shell of stars rotated, which required a special design and connecting material that was invisible and strong enough to hold them all in place. Tycho was not the first to encounter this demand on the imagination, because Isidore of Seville (c.560-636) also believed that the stars of the bounded Universe were at different distances.

Tycho fixed the Earth and allowed his shell of stars to turn daily about it, but a contemporary, Reymers Bär (1550-1599), suggested the opposite. Tycho had a prolonged struggle with him over priorities, and had disagreements with Duncan Liddell (1561-1613) as well, who claimed in private to have invented the model even though he still graciously called it Tychonic. Paul Wittig (c.1546-1586) was a mathematician who once served as Tycho's assistant and whom Tycho hoped would work out the details of his model, but this never happened and his hybrid failed to serve as a significant alternative to other models.

In the last years of his life, Tycho sought help from Kepler, but was again disappointed. Kepler's mathematical knowledge was of little help to Tycho, but Tycho's observational data were a great help to Kepler. Using them, Kepler discovered the three empirical "laws" of planetary motion that bear his name. He published the first two in 1609 in *Astronomia Nova* (The New Astronomy) and the third a decade later in *Harmonices Mundi* (The Harmony of the Worlds). From the time of Pythagoras, the music of the spheres has been a metaphor for an aesthetic appreciation of the Cosmos and Kepler's three mathematical relationships qualify eminently. Sadly, Tycho died in 1601 in ignorance of the harmony of the worlds that his life's work had helped bring about.

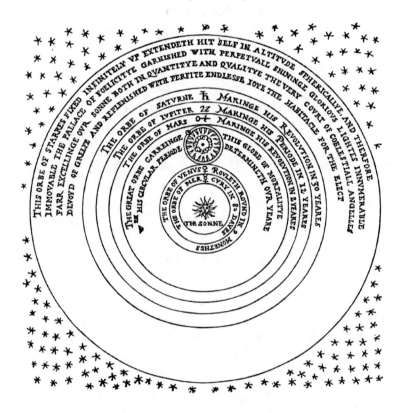

Figure 6: The unbounded model of the Universe according to Thomas Digges. (From *A Perfit Description of the Caelestiall Orbes*.)

Thomas Digges.

In 1576, twenty years after heliocentricism had started to take root in England, Thomas Digges proposed a model that imbedded a Solar System in an infinite Universe of stars (see Figure 6). He entitled his essay *A Perfit Description of the Caelestiall Orbes according to the most auncient doctrine of the Pythagoreans, latelye reuiued by Copernicus and by Geometricall Demonstrations approued*, and published it in a popular almanac founded by his father. This short essay advocates all essential features of the New Philosophy. Just as his contemporary, Tycho Brahe, argued for the destruction of the planetary spheres, so Thomas Digges advocated destruction of the sphere that held the stars, except that in Digges' case, the advocacy was a far more serious matter because, in an unbounded Universe, there is no end to secular inquiry.

A Recurrent Theme.

A recurrent theme in the development of science is that novelty encounters resistance when it flies in the face of cultural or religious norms. Ridicule and persecution often occur in the early and most vulnerable stages of paradigmatic shifts. We need look no further than Pythagoras who, as a young man, fled the tyrant who ruled Samos, the island where he grew up. After traveling widely, Pythagoras founded a school in Italy where he and his followers examined religious beliefs and developed novel opinions about the heavens. Persecution drove the school out of existence and Pythagoras died when hostile townsfolk burned the house he was in..

The persecution presaged the tribulations of some of his successors. For example, in the fifth century BC, residents of Athens regarded as blasphemous the idea that the Earth's shadow causes lunar eclipses. Anaxagoras suffered exile for believing that the Sun is a mass of molten iron even larger than the Peloponnesus. The philosopher, Cleanthes, asserted that it was the duty of all Greeks to indict Aristarchus on the charge of putting the Earth in motion.

Heraclides Ponticus advocated a rotating Earth and a Universe of infinite extent, which greatly offended the last major Greek philosopher, Proclus (*c.*410-485), who denied that Heraclides was even a student of

Plato. Not content merely to assail the theories of Heraclides, detractors mocked his name and his garb, gait and girth.

Copernicus kept details of his theory hidden for nearly 36 years because of the fear that he felt. Two years after *Commentariolus* had appeared he was ridiculed on stage near Frauenberg. In *Table Talks* of 1539, Martin Luther raised objections, calling Copernicus a fool for wanting to reverse the entire course of astronomy. Luther's chief follower, Melanchthon (Philipp Schwarzerd, 1497-1560), a Professor at Wittenberg, raised objections as well; in *Initiae Doctrinae Physicae* of 1549-50 published in Wittenberg, he called Copernicus a fool who was copying Aristarchus for purposes of self-promotion. He deemed heliocentricism "absurd" and hinted that a wise government should not tolerate Copernicanism.

Galileo knew of the scorn heaped upon Copernicus and, in 1597 in a letter to Kepler, he wrote that he had decided not to support him openly. In 1610, Galileo's *Sidereus Nuncius* (The Sidereal Messenger) itself was widely scorned and later, censors delayed publication of his *Due Nuoue Scienze* (Two New Sciences). Initially, Galileo had the support of Church authorities but the novelty of his thinking and his intolerance of arguments by his detractors resulted in an indictment for heresy. He was brought before the Inquisition and convicted, and was lucky to escape with nothing worse than house arrest and some mandatory incantations.

By 1616, Copernican theory had received the official censure of the Church. *De Revolutionibus* appeared on the *Index Librorum Prohibitorum* (Index of Prohibited Books) along with works of Galileo and Kepler, where they kept company with other works deemed needful of correction or incineration. In retrospect, a clash between theological inerrancy and the New Astronomy seems inevitable as both strived to accommodate one another at the fringes of their respective spheres of interest.

England had her share of troubles. For example, from 1558 to about 1578, John Dee was the leading light of English science and Queen Elizabeth (1533-1603) much admired him, yet Dee's enemies sought incriminating evidence against him that would suit their beliefs. In 1583, a hostile mob ransacked his house and destroyed his library and his scientific accoutrements, although, thoughtfully, they left his dwelling intact. Dee had served the queen well, and when news reached her of this event, she was outraged and promptly secured his house against further attack. These and related events most likely attracted the attention of churchmen and intellectuals across Europe. This was a time when intellectuals could not rest easy.

Gettings (96) describes "a tradition in esoteric history that whenever a new culture is embryonic in the womb of an older one, or when an esoteric school recognizes that a culture has served its purposes and is coming to an end, then a major work of art is created in dedication, as an outward sign for future ages." Further, "whatever its external artistic form, it encapsulates, in entirely esoteric principles, a summary of what has gone before, and what is to come. All the great esoteric artists, from Dante to Shakespeare, from Milton to Blake, have recognized this primal function of their art." One wonders whether any poets in the early modern age foresaw the impending upheaval in worldview.

CHAPTER 2:
POETRY AND COSMOLOGY

Poetry, the mistress of all discovery.
Ben Jonson

By the turn of the seventeenth century, the hypothesis of a rotating and revolving Earth had been in print for nearly sixty years, but poets were slow to adjust. Gabriel Harvey (1545?-1630) was an early enthusiast of contemporary astronomical learning and he attributed the deficiency to a general lack of scientific learning. He names several poets from the past who produced notable astronomical descriptions, including Ovid (43 BC - 18 AD), Seneca, Lucan (39 AD - 65 AD), Marcus Manilius (*fl.*10 AD), Francesco Petrarca (1304-1374), Dante, Girolamo Fracastoro (1483-1553) and Pietro Angelo Manzoli (*c.*1500-1543), better known as Palingenius. Harvey writes that poets must be more than superficial humanists and he names Geoffrey Chaucer (*c.*1340-1400) and John Lydgate (*c.*1370-*c.*1450) as examples of "exquisite artists, & curious vniuersal schollers." In the sixteenth century, it seems that no English poet fully supported the New Philosophy.

Donne.

John Donne (1572-1631), the "Copernicus in poetry," began writing about 1590 when Copernicanism had begun to make inroads in England. He addresses issues raised by the New Philosophy and, in 1623 in *Devotions upon Emergent Occasions*, he calls himself "a new Argument of the new Philosophie." His satirical *Ignatius his Conclave* of 1610 is a "daring piece of invective and scurrility" (Hardy 123-4) in which Lucifer questions the credentials of Copernicus at the gates of Hell. He advances a physical argument for a moving Earth by describing dancing newlyweds whose feet never return to the same place. In *A Valediction: forbidding mourning*, Donne recognizes that the circle is a symbol of perfection but notes that the circular motion of Earth brings "harms and fears." If the Earth rotated, loose objects would, he thought, lag and fly away in a direction opposite to the motion and, as a result, "Moving of th'earth brings harms and fears," whereas "trepidation of the spheres ... is innocent." The words "Moving of th'earth" and "trepidation" are ambiguous, however. The customary explanation is that they refer to earthquakes and the fears of defenseless travelers, but Donne may allude to the astronomical phenomenon of Trepidation.

In *Verse-Letter to the Countesse of Bedford* of 1611, he writes that the "new philosophy arrests the Sunne, / And bids the passive earth about it runne." In the same year in an oft-quoted passage from *The First Anniversary, An Anatomy of the World*, he says that people are confused:

> And new philosophy calls all in doubt,
> The element of fire is quite put out.
> The sun is lost, and th'earth, and no man's wit
> Can well direct him where to look for it ...

Donne refers specifically to two vital parts of the New Philosophy, the heliocentric model of the planetary system and the demise of the ancient element "Fire."

Donne does not embrace the New Philosophy wholeheartedly, however. *The Sunne Rising* is a humorous love poem and a reaction to the Copernican dominance of the Sun in which, in a counter-Copernican maneuver, he strips the Sun of its majesty in order to assert human centrality in the Cosmos. He expresses concern for moral virtue in the event that humankind were no longer to reside at the hub of creation. Like many of his contemporaries both English and Continental, Donne was "neither a champion nor an opponent of the new scientific theories" which, apparently, he "neither believed nor disbelieved" since their truth or falsity was "ultimately irrelevant to his true concern," which was the soul's salvation (Warnke 16).

Donne's anthropocentricism, his "human-centeredness," suggests indifference toward the scientific aspects of the New Philosophy. His poems "show an appreciation of Copernican doctrine ... enriched by allusions to the Old Philosophy," and as a result, "we cannot associate him with specific and well-defined currents of opinion or schools of thought" (Coffin *Philosophy* 97, 294). Donne "presents himself not as a systematic philosopher or theologian, but as a poet or a preacher who, having committed himself neither to the Old or the New Philosophy, felt free ... to make any figurative use whatever of the scientific matter in which he was versed" (Matsuura 1). In short, he "chooses the philosophy that illustrates what he wants to say" (Gorton).

Other Poets.

Other English poets who knew of the New Philosophy include Fulke Greville (1554-1628), George Chapman (1559?-1634), Christopher Marlowe (1564-1593), Thomas Nashe (1567-1601) and John Davies (1569-1626), but they show no great interest in the subject and make only fleeting references to it. Greville's ideas seem progressive enough although their chronology is hard to pin down. In *Of Monarchy* he acknowledges the Earth's motion and the Obliquity of the Ecliptic and, in *A Treatie of Humane Learning*, he writes that astronomy cannot decide whether we should believe what we see. Greville distinguishes reality and appearance by drawing attention to the inadequacy of visual perception in deciding matters of cosmology and says that humankind must not scorn science. Chapman's *Tears of Peace* of 1609 refers to disbelief in the rotation of the sphere of the stars and also notes the resulting state of uncertainty.

Marlowe was a Renaissance skeptic who knew of the heliocentric hypothesis but chose to stick to bounded geocentricism, which he presents as a simple arrangement of concentric spheres. He alludes frequently to basic concepts in astronomy without cluttering his writing with detail. In *2 Tamburlaine*, he writes that, "the massy substance of the earth" might "quiver about the axle-tree of heaven." By "axle-tree" he could mean the Earth's axis either of rotation or of revolution but, like Donne, he might simply refer to earthquakes. The year 1592, saw the first performance of *Doctor Faustus* in which Mephistopheles converses with one John Faustus who sells his soul to the devil. Marlowe named Faustus for the legendary Johann Faust (*d.*1541), a necromancer and astrologer. The conversation contains a concise description of bounded geocentricism and one cannot help but wonder whether Marlowe equates geocentricism with the devil incarnate.

Nashe was renowned for satire and prejudice. His penchant for ridicule resulted in a quarrel with Gabriel Harvey and came to account for more of Nashe's writings than any other topic. Nashe championed

Aristotle whereas Harvey supported Ramus, and their assaults ceased only after books by both were banned. However, the occasion of the two residing by chance one night in adjacent rooms at an inn provided the incentive for Nashe to resume his invective. In 1596, in *Have with You to Saffron-walden: or Gabriell Harveys Hunt is vp*, Nashe ridicules Gabriel's support for the new World view and writes that, when Harvey studies heliocentricism, he becomes so rapt that "hee would remaine three dayes and neither eate nor drinke." Like the detractors of Heraclides Ponticus two thousand years earlier, Nashe ridicules Gabriel's physique.

John Davies keeps an open mind about the New Astronomy. In 1596 in *Orchestra or A Poem of Daۥncing*, he writes that the Earth stands forever still, but he adds, parenthetically, that some learned wits say that the heavens are still and the Earth rotates. Robert Burton (1577-1640), in a vast wide-ranging treatise, *The Anatomy of Melancholy* of 1620 and later, devotes considerable commentary to the various world models of the time, saying that the New Philosophers are disparaging the concept of Fire as well as the idea of Heaven lying above the Firmament. He deals harshly with the wheels, spheres, and cogs of geocentric machinery, which he describes as monstrous orbs that are so absurd and ridiculous that no one could possibly think that there should be so many circles, "like subordinate wheels in a clock, all impenetrable and hard" which Ptolemy and his followers "add and subtract at their pleasure."

In *An Hymne of Heavenly Beauty* of 1595-6, Edmund Spenser (*c.*1552-1599) describes the basic geocentric model. He starts at Earth and continues outward to the sphere of the stars and thence to the Creator, "First the Earth ... / And last, that mightie shining christall wall, / Wherewith he hath encompa·sséd this All." The crystal wall is the boundary that separates natural and "supernatural" space. Between the Earth and the stars lies The House of Blessed Gods, which is the abode of the Ancient Planets. These shine by compressed Fire, but their brightnesses pale by comparison with the light beyond the starry welkin, which is unbounded, uncorrupt and infinite in largeness and light.

Milton wonders whether the Sun is the center of the Universe, and whether the Earth has the three independent motions. In *Paradise Lost* of 1667, he shows that he knows about the alleged phenomenon of Trepidation for he speaks of "that crystalline sphere whose balance weighs / The trepidation talk'd, and that first mov'd." He writes skeptically of it, as he should, since the phenomenon had long been consigned to history books. After the young Cambridge graduate, Isaac Newton (1642-1727), had experienced his "wonderful years" 1664-1666, Milton still seeks safety in bounded geocentricity.

Galaxia.

The Milky Way is a nebulous band of light stretching across the sky and takes its name from the Greek "gala," meaning "milk." From this root, the words galaxia and galaxy are formed. The Milky Way is most prominent in southern skies, but it is striking enough when seen from dark sites in the north. Telescopes reveal that this pearly band comprises, in part, a myriad of distinct stars so crowded together along the line of sight that, when viewed by the naked eye, they blend and take on a milky appearance.

There are many classical references to the Milky Way. In *Commentary on the Dream of Scipio,* Macrobius (*fl.*400 AD) cites the atomist Democritus as one of several who hold opinions on the phenomenon. Democritus and his followers explain the Milky Way as a luminescence due to the blending of many faint stars crowded together. Anaxagoras and Aristotle held similar views. These early thinkers believed that the stars of the Milky Way are visible because the shadow of the Earth falls them, rendering their light more noticeable by contrast with regions that the Sun illuminates.

About the turn of the twelfth century, the Oxford don, Robert Grosseteste (*c*.1170-1253) wrote that people knew that the Milky Way is an agglomeration of closely spaced stars. His successor, Roger Bacon (*c*.1214-1294?), held the same opinion. Both worked on mirrors and lenses and, using them, may have discovered evidence supporting their contention (see Chapter 4).

Bartholomaeus Anglicus (*fl.* 13th century) re-asserted the starry nature of the Milky Way. In his encyclopedic compendium of biblical and Aristotelian knowledge, *De Proprietatibus Rerum* of about 1245, he writes, "And in the place where *Galaxia* is seene, be many small stars and bright, and in those stars shineth that brightnesse. And therefore that place seemeth most bright with beames of light." Bartholomew falsely attributes this idea to Aristotle, but the important point is that, after 1495, an explanation for the nature of the Milky Way existed in a summation of knowledge whose translation became one of the first books published in English. By about 1582, the compendium had evolved into its final form, which Shakespeare used as one of his sources.

Chaucer too, associated the Milky Way with stars. He refers to the Galaxy in *House of Fame* of about 1384, "See yonder, loo, the Galoxie, / Whiche men clepeth the Melky Weye." In *Parlement of Foulys*, he makes clear that he regards the Milky Way as the "path" or way to Heaven and refers to Cicero who, in Book VI of *De Republica,* writes of a dream of Scipio Africanus. Scipio flies across the planetary spheres to the Milky Way and, as he looks around, he sees stars that "we never see from this country." These are probably stars of the Southern Hemisphere, but they could be stars too faint to see from Earth that had become visible owing to their newly acquired proximity.

In *Troilus and Cresyde*, Chaucer describes the path of Troilus' soul and speaks of the hollowness of the eighth sphere. North (*Chaucer* 29-32) argues that there is some question about Chaucer's numbering of the celestial spheres but, on reviewing the evidence, concludes that "hollowness" refers to an aperture from the seventh to the eighth sphere, which would render stars on the eighth sphere more easily visible. Chaucer's region of hollowness is the antithesis of the so-called Zone of Avoidance that played a part in work by Heber Curtis (1872-1942) and Harlow Shapley (1885-1972) on the discovery of so-called "extragalactic nebulae," which we know now are galaxies like our own Milky Way. Rather than being a zone of high opacity, Chaucer sees the Milky Way as a zone of transparency and thus as a pathway in physical space to the edge of supernatural space.

In 1582, Thomas Watson (1557?-1592) produced *Hekatompathia* in which a few lines of Sonnet 31 refer to the stars of Galaxia, and Sebastian Verro (*fl.*1581) made similar remarks the year before (Altschuler and Jansen). Watson's four lines reflect the state of knowledge expressed in the encyclopedic compendium of Bartholomaeus Anglicus and the works of Grosseteste and Roger Bacon. Thomas Watson was in Paris at the same time as Giordano Bruno, who was then most likely developing his own ideas about theological and physical infinities. It is likely that the two met there prior to Watson penning his Sonnet 31, particularly since Watson was an ardent admirer of Italian literature and Bruno, in turn, valued the opinions of Democritus on the nature of the stars. By this time, Lambert Daneau (1530-1596) had discussed atomism in *The Wonderfull Woorkmanship of the World*, which appeared in English translation in 1578.

Shakespeare.

The Bard lived at a time of great change, when the medieval world-order was under review and wider vistas were opening up. Mowat and Werstine (*Lost* pp. xxxiii-xxxv) write that Shakespeare's productive years "were among the most exciting in English history." The medieval World view was in crisis, yet no one can point with any conviction to the places in the Canon where Shakespeare expresses knowledge or appreciation of emergent cosmologies and the New Philosophy.

Simpson (41-2, 42n1) remarks that all the great Elizabethan and Jacobean dramatists "seem to have been quite unmoved by the achievements of science." Along with Ben Jonson (1572-1637), John Webster (1580?-1625?), Francis Beaumont (1584?-1616) and John Fletcher (1579-1625), he mentions William Shakespeare. McAlindon (4) believes that "Shakespeare's understanding of nature was fundamentally traditional" and that, even after the new science had begun to change the picture of the Universe and of humankind's relation to it, "there are no signs of this revolution in his work." Bevington (p. xxvi) writes, "All major poets of the Renaissance, including Shakespeare, Spenser, and Milton," adhered to the standard Ptolemaic World view. Marlowe set the literary stage for Shakespeare, and both seem devoted to geocentric orthodoxy. Asimov (I, 25) writes (emphasis his), "Shakespeare does *not* ... take the advanced position of agreeing with Copernicus." Concerning the radical revision of the location of stars, Hotson (*Appoint* 123) writes, "One must admit that among Shakespeare's myriad minds, there was not the mind ready to kindle to the truth of [Thomas] Digges' vast vision [of an infinite Universe]."

Nevertheless, as we have seen, the New Philosophy did attract a modicum of attention from poets, and some believe that the Bard, too, was not fully committed to the Old Astronomy. Bevington (444) writes that *Troilus and Cressida* and *Hamlet* concern the "loss of an assured sense of philosophical reliance on the medieval hierarchies of the old

Ptolemaic earth-centered cosmos." As Donnes puts it, the Universe is "all in pieces, all coherence gone." *Troilus and Cressida* was written probably about 1601 or 1602, about the same time as *Hamlet*, but for some reason was not printed until 1609 and then in two versions. In one version, printers altered the title page and added an unsigned epistle entitled "A Never Writer, to an Ever Reader. News," where, presumably the "never writer" refers to the publisher. The note alludes to the possibility of an English Inquisition like the one on the Continent that ferreted out heretical documents and beliefs. The epistle ends with *Vale,* meaning "farewell," which is also how Osiander ends his gratuitous preface to *De Revolutionibus*.

Ulysses is a Greek commander and political operative who appears to uphold traditional virtues of rank and order, and his "degree" speech in *Troilus and Cressida* contains the only passage that might indicate an acceptance of heliocentricity:

> The heavens themselves, the planets, and this centre
> Observe degree, priority, and place ...
> And therefore is the glorious planet Sol
> In noble eminence enthroned and sphered
> Amidst the other ...

This may mean that the Sun is at the actual physical center of the planets, but the lines are ambiguous because, in the sequence (A) of Chapter 1, the Sun has three Ancient Planets on either side of it and, in that sense, is in the "middle" of the others. Contemporary writing supports the former interpretation.

In *A Perfit Description*, Digges refers to the planets and states in a paragraph of one line, "In the myddest of all is the Sunne." Digges likens the Sun to a reigning monarch, "the Sunne ... like a king in the middest of all raigneth and geeueth lawes of motion to ye rest, sphaerically dispearsing his glorious beams of light through al this sacred Caelestial Temple." Digges repeats the metaphor, "thus doth the Sun like a king sitting in his throne govern his courts of inferior powers." He notes that some call the Sun "the Ruler of the worlde" that gives "laws of motion" to the rest of the Solar System.

The promotion of the Sun to the seat of royalty occurs also in the Copernican treatise. "Nothing is more repugnant to the order of the whole and to the form of the world than for anything to be outside of its place," he writes as he raises the stature of the Sun and puts planets in their proper places. That done, Copernicus writes, "In the center of all rests the sun ... And so the sun, as if resting on a kingly throne, governs the family of stars [i.e., planets] which wheel around." The uncanny resemblance between the words of Shakespeare, Copernicus, and Digges suggests that Shakespeare enthrones Sol at the physical center of the planetary orbits and perhaps the Bard is not as backward as we think.

Hierarchy is at issue. Plato's Principle of Plenitude, which Arthur Lovejoy (1873-1962) re-phrased and called the Great Chain of Being, argues that divine creativity necessarily produces all possible existents, leading to a full and continuous array of beings. God presides at the pinnacle of creation, followed by physical existents of ever-diminishing stature. Monarchs and emperors are closest to God, and lesser mortals fall below them in an ever-increasing cascade of categories. In the physical Universe, the Earth occupies the central nethermost slot and above and around it are arrayed the Ancient Planets ascending in order toward Heaven and the Creator. Everything has a place and, in the best of all worlds, everything is in its place. Ulysses upholds the concept of orderliness but foresees an end to the medieval and political world order. Ulysses has cosmic hierarchies in mind for he uses planetary images to warn of the impending chaos, "when the planets / In evil mixture to disorder wander."

Of all plays in the Canon, *Troilus and Cressida* has the highest incidence of the names of the five retrogressing planets. The meanderings of these troublesome tramps suggest that they do not know their proper place in the grand scheme of things. Upsetting the divine order by removing the Earth from the center and placing the Sun there instead is an invitation to disaster and Ulysses warns that, without priority and place, discord will follow. He refers to the Pythagorean belief that music symbolizes the harmony of a finely tuned Universe. The planets take different times to undergo a circuit of the Earth, just as the taut strings of the lyre vibrate with different frequencies. By analogy, the Music of the Spheres is a heavenly harmony audible to the imaginative ears of initiated Pythagoreans, as Lorenzo tells Jessica in *The Merchant of Venice*. Lorenzo says that all stars sing like angels, so a planet that no longer toes the canonical line is like a musical string that is untuned, as when, according to Troilus, "The bonds of heaven are slipped, dissolved, and loosed." One is tempted to believe that Shakespeare's *Troilus and Cressida* advocates a return to the Pythagorean ideals that undergird the new World view, almost as if it were a precursor to a more thoroughgoing exposition.

The question of hierarchy had implications for Elizabethans since an unstable cosmic state could reflect unfavorably on the political state. Elizabeth lived under constant threat to her life, beginning when she became a rallying point for anti-Marian Protestants. After she ascended to the throne, Pope Pius V (1504-1572) issued a decree, *Regnans in Excelsis*, excommunicating her. Among her many transgressions was that she introduced books of heretical content into her realm. The wording of the bull was ambiguous, and some took it as an invitation to assassinate her. Often in politics as in physics, action and reaction are equal and opposite, and Huguenot refugees from the Saint Bartholomew's Day massacre carried the idea a step further and urged the assassination of tyrants of any kind. At the turn of the seventeenth century, the aging Queen Elizabeth had not named a successor and it was not a time for poets to write plainly of new hierarchies.

For this reason, perhaps, Shakespeare's references to the New Astronomy are ambiguous and he may well have opted for discretion by limiting himself to simple commentary on the heavens. For example, he adheres to the standard doctrine of crystalline spheres supporting Ancient Planets placed in order between Earth and the Firmament. Beyond the quintessential sphere of the heavens lies the abode of "that supernal judge that stirs good thoughts / In any breast of strong authority" (*King John*). Here is Shakespeare's only use of the word "supernal," meaning "celestial," "heavenly," or "existing or dwelling in the heavens," and when it modifies "judge," it means God. His frequent mention of "heaven" reflects a need for dramatic expression. He makes the simplest connections between Heaven, Earth, and the Ancient Planets, as when Prince Arthur leaps to his death exclaiming, "Heaven take my soul, and England keep my bones!" In *Troilus and Cressida,* we read of "the very center of the earth, / Drawing all things to it." This is consistent with Aristotelian-Thomist physics in which the base element Earth occupies the lowest state. From *Pericles*, we learn of the four elements Fire, Air, Water and Earth. Hamlet describes the sky as "this brave o'erhanging firmament, this majestical roof fretted with golden fire." Caesar says the sky is "painted with unnumber'd sparks" which "are all afire and every one doth shine." In *Coriolanus*, a messenger measures certainty by knowing that the Sun is Fire and in *Macbeth* the Thane of Glamis commands the stars to hide their fires. From *Antony and Cleopatra* we learn that the Sun and Moon are stuck between Earth and Firmament. Sunlight robs the oceans of water and the moon is an arrant thief because "her pale fire she snatches from the sun" (*Timon of Athens*).

Shakespeare refers often to the appearance of sunrise in the east and to the morning hours (*Sonnet* 7), "Lo! in the orient when the gracious light / Lifts up his burning head." The Sun moves along "his bright passage to the occident" and sets in the west (*Richard II*). A giant transparent crystalline orb holds each planet, like "yonder Venus in her glimmering sphere" (*A Midsummer Night's Dream*). The day and the year are empirical manifestations of the two basic motions of the Earth in the Copernican model, viz. its rotation on its axis and its revolution about the Sun, but the Bard does not relate this elementary empirical evidence to Copernican science. A calendrical day is merely the interval between successive appearances of the Sun above the horizon, and no year passes, "until the twelve celestial signs / Have brought about the annual reckoning" (*Love's Labour's Lost*). The year is marked also by the seasons during which daylight lost in winter is like a debt repaid by the long days of summer (*Timon of Athens*).

In *As You Like It*, the Bard abides the Mosaic time scale for the World, "The poor world is almost six thousand years old." James Ussher (1581-1656) studied Biblical chronology and estimated the time of creation of the world at 9:00 a.m. on October 23, 4004 BC. Counting each day of creation as a millennium (*2 Pet.* 3.8), it follows that, in the sixteenth century AD, the world was only a few centuries shy of "six thousand years old." In *Henry IV part 1*, Shakespeare goes on to warn that some day it will all end, "time that takes survey of all the world, / Must have a stop."

Shakespeare caters to common superstition by allowing transitory phenomena, like meteors, comets and eclipses, to presage changes in governance or state. Meteors that frighten the Fixed Stars of heaven forerun the death or fall of kings, and comets, importing change of time and states, scourge the bad revolting stars. The gods care little about us hoi polloi, because when beggars die, there are no comets seen, whereas the heavens themselves blaze forth the death of princes. Astrologers use the apparent motions of Sun, Moon, and planets against the backdrop of the Zodiac to predict the future, but Shakespeare ridicules astrology and purported astral influences, as when Edmund says of his father's superstitions, "This is the excellent foppery of the world, that when we are sick in fortune ... we make guilty ... the sun, the moon, and stars." The disavowal of *Sonnet 14* makes the point convincingly:

> Not from the stars do I my judgment pluck,
> And yet methinks I have astronomy –
> But not to tell of good or evil luck ...

The Bard's aversion to pedantry is evident in *Love's Labour's Lost* where he assails those who accept Aristotle as their ultimate authority:

> Study is like the heaven's glorious sun
> That will not be deep searched with saucy looks.
> Small have continual plodders ever won
> Save base authority from others' books.

Shakespeare's knowledge of eclipses is more than rudimentary (Levy "Eclipses") and the enthusiasm of his negations of superstition raise the question of whether the Bard knows more than he lets on. An ambiguous reference to Mars heightens suspicions further. In the sixteenth century, the motion of Mars was a famous problem because, of all the planets, its ephemeris was the most intractable. In 1609 in *Astronomia Nova*, Kepler stated that Mars had an elliptical orbit, but about eighteen years earlier in *Henry VI part 1*, Shakespeare wrote, "Mars his true moving, even as in the heavens, / So in the earth, to this day is not known." The passage is ambiguous to the extent that neither the geocentric nor the Copernican algorithms forecast Martian positions accurately.

Lodestar.

Before the second century BC, people believed that the Firmament rotated regularly about the Earth and, thus, that the directions of the Celestial Poles were fixed relative to the stars. In principle, after the discovery of Precession, no one could regard the Celestial Poles as "forever fixed" in the sky because, over time, Precession would cause the direction of the Poles to move relative to the stars. After Ptolemy, hardly anyone acknowledged it. Proclus was one of the last of any distinction to do so and then only to deny it. Passages in the Canon might lead us to conclude that Shakespeare is ignorant of the phenomenon (Allen 458; Roy), but counterarguments suggest otherwise.

In *Othello*, two provincial gentlemen of Cypress speak in a manner that, to this writer's ear, smacks of certainty stemming from ignorance. The Second Gentleman opines that the North Celestial Pole is a direction that is "ever fixed" in the sky. The direction of the North Celestial Pole to which the insular gentleman refers seems to him fixed and, if that gentleman were ignorant of the history of astronomical discovery, he would be none the wiser during his lifetime because the precessional cycle lasts 26,000 years. The legendary Methuselah lived for nearly a thousand years and might have noticed a change in the position of the North Celestial Pole, but only if he had a memory as long as his life. Perhaps Shakespeare is simply catering to common perceptions rather revealing ignorance of the phenomenon.

Plato and Cicero describe the apotheosis of the fixed and wandering stars and the belief that the stars represent the souls of the departed. The star nearest the North Celestial Pole commands attention because, night after night throughout a normal lifetime, it seems fixed in the sky. This cynosure has special significance to navigators because it serves as a beacon by which to determine the direction of north. One would think that the esteemed post-Hipparchian Emperor, Julius Caesar (102?-44 BC), would know this since generals need to know in what direction to lead their armies. Yet, in a dramatization that accords with history, Shakespeare allows Caesar to harbor ignorance of Precession. Caesar believes that he is special because he is "constant" like the North Pole Star of whose "true fixed and resting quality / There is no fellow in the firmament." He thinks he is entitled to reserve this star for his own soul. Caesar believes that only one person is constant and special and, he says, "I am he." Less than twenty lines later, he is dead. Caesar has achieved a "resting quality," although not of the sort he expected.

By killing off Caesar so suddenly, Shakespeare shows that the Emperor is not "constant" any more than a Cynosure remains always near the Pole. Eventually Caesar's soul will become just another star in the firmament because at no time does a hypothetical star lying in the precise direction of the Pole have a true-fixed and resting quality. Precession dashes Caesar's hope for eternal preeminence and the thesis that Shakespeare is ignorant of the phenomenon does not hold up.

It is unclear to which "Northern Star" Shakespeare refers. In *Julius Caesar,* A likely candidate is β Ursa Minoris, known to the Greeks as Polos, the Pole Star, but if a contemporary time obtains, the likely referent is α Ursa Minoris, which English navigators called *Stella Polaris*, the Steering Star. In *Sonnet 116*, Shakespeare refers to a navigational beacon that is most likely Polaris:

> Love ... is an ever-fixèd mark
> That looks on tempests and is never shaken;
> It is the star to every wandering bark,
> Whose worth's unknown, although his height be taken.

From 1551 to 1726, "height" means the elevation angle of a heavenly body above the horizon (*OED*). From 1585 to 1694, "height" is also geographic latitude (*OED*), a coincidence explained by simple geometry and by the fact that Claudius Ptolemy established the convention of measuring latitude from the Earth's Equator and putting the direction of north at the "top" of a map. The altitude angle, or "height," of the North Celestial Pole above the horizon equals the geographical latitude angle, or "height," of the observer "above" the equator, which is important to navigators aboard wandering barks who want to know their latitude. The "mark" in the sky is, supposedly, forever indicated by a star whose altitude angle is measured as stated, but the difficulty is that, for some reason, its "worth's unknown."

Uncertainty in the measurement of the altitude of *Stella Polaris* could arise from a correction that the navigator might make to account for the refraction of light by the Earth's atmosphere. This effect was known at least from the time of Ptolemy and, in principle, navigators could make it, although at latitudes of interest to Shakespeare this effect would amount to a relatively small navigational error. Uncertainty could arise also from the likelihood that, at the time of observation, Polaris might not lie exactly in the direction of the North Celestial

Pole, but above or below it or somewhere in between, resulting in a derived value for latitude that depended on the time of measurement. If necessary, a conscientious navigator could correct for the effect as well. If Shakespeare had a longer time scale in mind, the "height" of the ever-fixed mark has a measure of worth that is unknown because of confusion about Precession.

Other Sciences.

When it comes to sciences other than astronomy, the Bard's writing is explicit. He appreciates the ecological balance of the Earth, "For naught so vile that on the earth doth live / But to the earth some special good doth give" (*Romeo and Juliet*). A geologist on an expedition to Antarctica "pass[ed] the time of day" during the "long winter night" by reading all of Shakespeare's plays and was astounded at the frequency and accuracy of the poet's geological allusions (Gould p. xx). In *Henry IV part 2*, we read of slow changes that level mountains and cause continents to melt into the sea. In physiology, Brutus speaks of "ruddy drops / That visit my sad heart" (*Julius Caesar*), and of "rivers of blood" that service the heart and the brain (*Coriolanus*).

Topics like geology and physiology would scarcely inflame the enemies of the English state. By contrast, the New Astronomy threatened the revered pyramid of Plato's hierarchy, and, even worse, carried to extremes, could impinge upon the space reserved for the Almighty. Such threats to accepted doctrine would engender hostility both domestically and abroad. It seems that Shakespeare has a thorough grasp of scientific matters but his indifference to the Earth-shaking revolutions of the sixteenth century and their antecedents in classical antiquity prompt further inquiry.

CHAPTER 3: THE IDEA OF INFINITE SPACE

> In order for the oppressed to be able to wage the struggle for their liberation, they must perceive the reality of oppression not as a closed world from which there is no exit, but as a limiting situation which they can transform.
>
> Paulo Friere

In the late sixteenth century, academies of higher learning acknowledged the existence of the Copernican World view but did not take it seriously. Occasionally, faculty offered a university course in astronomy at an advanced level, as in 1570, two years before the New Star of 1572, when Henry Savile (1549-1622) delivered a series of special lectures on Ptolemaic astronomy. Only in 1619, with the establishment of Savilian Professorships in geometry and astronomy at Oxford, did heliocentricism achieve academic recognition among leading English universities. This was nine years after Galileo had made his epochal telescopic discoveries and ten years after Kepler had announced his first two empirical laws of planetary motion.

The backwardness of the leading English universities at the time is evident by the absence of chairs in mathematics and by the short university tenures of leading scientists like Robert Recorde, John Dee, Thomas Hood (*fl.*1582-1598), and William Gilbert (*c.*1540-1603). Despite their independence, Recorde and Dee became leaders in mathematical studies in England. For twenty years following Recorde's death in 1558, Dee was the authority in matters of science and geography and his house near London became a mecca for those seeking such knowledge. Science in England flourished thanks to the initiative of individuals who distanced themselves from the chief academic centers of learning and whose creativity matched the depth of their pockets. London and neighboring counties soon became the center of independent learning and, in 1596, the newly founded Gresham College created professorships in geometry and astronomy.

Science in the Vernacular.

Shortly after 1517, John Rastell (1475-1536), brother-in-law to Thomas More (1478-1535), advocated the use of English in scientific writing. Scholars began to write books on science from which students acquired most of their knowledge, and almost all were in English. Writers attempted to use English roots for new technical terms, but the practice proved cumbersome and they resorted to Latin roots, the quintessential "straight line" being one of the few surviving neologisms. In 1573, Thomas Digges wrote *Alae seu Scalae Mathematicae* in Latin in order to promote accessibility of the work to readers on the Continent, but three years later, he too published in English and, in 1591 in the second edition of *Pantometria,* he renounced Latin for English once and for all.

For his part, Shakespeare scorns effete linguists, like Armado ("a congruent epitheton") and Polonius ("tragical-comical-historical-pastoral"), and he satirizes those who choose Latin over English, like Holofernes ("*honorificabilitudinitatibus*"). If perchance, Shakespeare had had the same interest in contemporary cosmology as those who promoted English over Latin, he would have had more than a cursory relationship with members of the new guard. Thomas Digges is a likely candidate for his contact in astronomy, especially since his father, Leonard Digges (*c.*1521-*c.*1572), invented the world's first two-element optical magnifier.

The Digges Family.

Leonard Digges belonged to an ancient and prominent Kentish family with roots traceable to 1254. He attended Oxford but took no degree. Like Newton in the seventeenth century, he was both an able mathematician and a keen experimentalist. His wealth and status permitted him to indulge his passion for knowledge unimpeded by scholastic culture, but his career nearly ended when, in 1554, he was an unlikely participant in an insurrection against Queen Mary I (1516-1558). She had ascended the throne in the previous year and announced plans to marry Philip II (1527-1598) of Spain. Philip was a religious fanatic and the proposed marriage did not sit well in Protestant Kent and neighboring counties. Thomas Wyatt (d.1554), the son and namesake of the Kentish poet, led a rebellion. It failed and Wyatt was captured and hanged. Leonard Digges was one of nearly 500 captives who were convicted of treason. His execution was scheduled for sometime after February 24, 1554, but an act of the Privy Council reprieved him. Cool heads must have prevailed because, in the end, only seventy-five of the rebels were executed. On April 1, 1554, Leonard received a pardon, along with many of his comrades-in-arms and, although history records the identity of most of the interceders, there is no such record in Leonard's case. Lord Clinton (1512-1585) had played a key role in suppressing the rebellion and had a reputation for diplomacy and practical understanding. He had an uncanny ability to serve monarchs regardless of religious persuasion and many believe he was the one who arranged the pardon.

On May 31, 1554, Leonard Digges entered into a monetary obligation for the redemption of his movables and, on 20 February 1555, he entered a second recognizance for the redemption of his lands, which he finished paying on May 7, 1558. In an age when state security demanded suppression of radicals, the survival of Leonard Digges was a remarkable feat. Having escaped the gallows, he resumed production of an almanac entitled *A Prognostication of Right Good effect* whose earliest known edition dates to 1553. In 1556, he renamed it *A Prognostication Euerlasting*. The almanac was extremely successful and served Leonard in the same way as *Poor Richard's Almanac* served Benjamin Franklin (1706-1790), who wrote in 1757 that he accrued some "solid pudding" from its sale.

Through his life and works, Leonard repaid the English state handsomely for its forbearance in keeping his body and soul together. Since Leonard disdained university curricula, he had John Dee tutor his son, Thomas, privately. Dee's influence was so strong that Thomas became an accomplished mathematician and eventually eclipsed his tutor's reputation. Thomas was active in politics as well and, from 1572 to 1583, was a Member of Parliament for Wallingford and after that, for Southampton, at which time his patron Robert Dudley (1532?-1588), Earl of Leicester and godfather to his son Dudley Digges (1583-1639), was instrumental in appointing him muster-master of the English forces in the Netherlands.

Thomas had studied the military history of ancient Rome and Greece and railed against what he perceived as the inadequacy and corruption of the English army. Consequently, his army pay was withheld and, in 1590, he had to petition the queen's chief advisor, William Cecil, Lord Burghley (1520-1598), for redress. Critics faulted Thomas for writing about military matters without combat experience, but opinion had it that he would have given a good account of himself in battle if the opportunity had presented itself. For example, he took to sea for fifteen weeks to demonstrate the correctness of his theories of navigation.

Thomas put his mathematical skills to practical use by writing also on parallax, geometry, military science, and ballistics. He was one of the first to demonstrate compound motion by dropping objects from the top of the mast of a moving ship, showing that it landed at the base of the mast and not somewhere aft of it. He knew that false methodologies prospered at the time and he knew of their affect on the two seemingly disparate fields of natural philosophy and politics. In a preface to a tract *Political Discourses upon Truth and Lying* by Edward Hoby (1560-1617), Thomas warns of faulty political philosophy. In natural philosophy, he was especially critical of Aristotle and, for that reason among others, he admired Pietro Angelo Manzoli, better known by his pen name Marcellus Palingenius Stellatus, or, simply, Palingenius, the Stellified Poet.

Palingenius.

In 1531, Palingenius wrote *Zodiacus Vitae* (The Zodiac of Life) in twelve books, each named for a Zodiacal constellation. The poem was a compendium of contemporary knowledge of which Book XI, subtitled *Aquarius*, dealt with astronomy. According to Gabriel Harvey, Thomas Digges memorized *Aquarius* by heart and was wont to repeat it often. By 1565, the plain-style poet, Barnaby Googe (1540-1594), had finished translating *Zodiacus Vitae*, which became popular in England and helped pave the way for challenges to the Old Astronomy. According to Schoenbaum (*Life* 69), in third form at school in Stratford, William Shakspere would have encountered Palingenius' compendium and thus have learned that all the world's a stage.

Palingenius is committed to geocentric orthodoxy and the finitude of the material world, but is, nevertheless, a free and original thinker. In *Aquarius*, he lifts the reader from the mundane world toward the heavens and, like his predecessors, asserts that God's heaven is associated with "infinity." He adopts the neo-Platonic view of a finite physical World imbedded in space of limitless extent that is flooded with pure light. He asserts that some stars are bigger than the Earth and only seem small because they are far away.

Palingenius assumes that the energy radiated by a star depends on its size. This is true up to a point, for the larger the star, the greater the surface area from which it can radiate and, therefore, the more light it emits. Palingenius says that the apparent brightness of stars depends also on distance d from the observer. "For everything beside," he writes, "The farther it is from our eyes, the less in sight is spied, / And do deceive the lookers on."

When an observer sees a star, his eye measures the flux of radiant energy entering his pupil and this depends on d and the power or luminosity L of the star. Palingenius knows that stellar appearances are not reality because he understands, at least qualitatively, what is today known as the inverse square law of flux from a small and/or distant source, $F = L/d^2$. This is equivalent to the distance modulus equation familiar to observers, $m-M = 5 \log d - 5$, where absolute and apparent magnitudes M and m are measures of L and F. Palingenius fixes all stellar properties ("for everything beside") and states that stars vanish from view if d is large enough. He says that onlookers are fooled into thinking that they are looking at stars of different L at the same distance d, when they could be looking at stars with the same L at different d.

In a purpose-driven Universe where all existents serve a function, would it not be an extraordinary coincidence if, in order to serve their purpose, all stars were visible precisely at or above the visual limit of the human eye? After all, if some stars were invisible, we would not know of them and they would serve no purpose. The situation is doubly difficult to accept in a Universe of Palingenius' design where, in order for stars at ever-increasing distances to have detectable fluxes F, they would need to have larger luminosities, L, in order to remain visible. Instead, Palingenius allows stars to lie at arbitrarily large distances so that, with the implicit assumption that stars have a finite power output, some must be invisible to the naked eye.

Palingenius makes it easy for devotees of Aristotelianism to take umbrage at ancient philosophers:

> What store of fond Foolosophers, and such as hunt for praise
> The earth brings forth, it is not good to credit all he says,

> Though great his estimation be in mouths of many men,
> Though many Reams of Paper he hath scribbled with his pen.
> For famous men do oftentimes make great and famous lies,
> And often men do miss the truth though they be [n]ever so wise.
> Therefore must reason first be sought, for in such doubtful things,
> More credit reason ought to have, than men's imaginings;
> For such are often proved false.

His criticism of Aristotle is withering:

> Whatsoever *Aristotle* saith, or any of them all,
> I pass not for, since from the truth they many times do fall.
> Oft prudent, grave, and famous men, in errors chance to slide,
> And many wits with them deceive when they themselves go wide.

One wonders whether Palingenius perceived the danger he was in. In 1558, *Zodiacus Vitae* appeared on the Index of Prohibited Books but, by then, Palingenius had been dead for fifteen years and did not feel the heat at the torching of his mortal remains.

A Perfect Description.

The finitude of the material Universe was a subject of vigorous debate during the Renaissance. The real break with the past occurred when bolder astronomers entertained the possibility of an infinite Universe, and Thomas Digges was the first post-Copernican to do so (see Figure 6). In 1576, thirty-three years after the death of Copernicus and five years after the death of his father, Thomas opined that editions of the almanac published before 1576 contained sundry faults, among which is a model of the World according to the doctrine of Ptolemy "whereunto all Vniursities (ledde thereto chiefly by the auctority of Aristotle) sithens have consented." He regards the Pythagorean School as the forerunners of the concepts of a rotating and revolving Earth and of stars populating infinite space. He flat-out contradicts Aristotle's advocacy of geocentricism and his idea that the Universe does not have even the potentiality of being infinite, although he blames Aristotle's disciples more than he does the revered philosopher.

Plutarch (46?-*c.*120) too, credits the Pythagoreans with the idea of multiple worlds. He writes that, "Heraclydes and the Pythagoreans

56

hold, that every Star is a world by it selfe, conteining an earth, an aire, and a skie, in an infinit celestiall nature," and Manilius writes that stars have different apparent magnitudes not necessarily because they have different luminosities but because they have different distances.

The ploy that Osiander perpetrated in the hope of protecting Copernicus from charges of heresy did not fool Thomas Digges. He writes that, "Copernicus mente not as some haue fondly excused him to deliuer these grounds of the Earthes mobility onely as Mathematical principles." He takes the heliocentric solution as an image of reality and imbeds it in an endless Universe strewn with stars. Like his hero, Palingenius, Digges predicts that stars become fainter with increasing distance until they eventually fade from view, giving the impression that we live inside an orb of stars. The following description occurs in the pictorial representation (see Figure 6):

> THIS ORBE OF STARRES FIXED INFINITELY VP EXTENDETH HIT SELF IN ALTITVDE SPHERICALLYE AND THEREFORE IMMOVABLE THE PALLACE OF FOELICITYE GARNISHED WITH PERPETVALL SHININGE GLORIVS LIGHTES INNVMERABLE FAR EXCELLINGE OVR SONNE BOTH IN QVANTITYE AND QVALITYE THE VERY COVRT OF COELESTIALL ANGELLES DEVOYD OF GREEFE AND REPLENISHED WITH PERFITE ENDLESSE JOYE THE HABITACLE FOR THE ELECT.

Copernicus had assumed that the Sun was stationary with respect to the stars, and Digges is silent on this matter. Neither knew what we know today, that the Sun is one of thousands of millions of stars that populate the Milky Way galaxy and are in motion relative to one another. Digges assumed that stars are the basic building blocks of the Universe and did not know that a hierarchy of larger entities exists. His *A Perfit Description* is just the first step toward establishing a new hierarchy for ordinary matter in the Universe.

Priority.

A Prognostication Everlasting appeared several times from 1576 to 1605 and, by the beginning of the seventeenth century, Digges' cartoon had become familiar to the English public, particularly those more inclined to look at pictures than to read text. By publishing his theory with its far-reaching consequences in a popular almanac, Digges perpetrated a ruse that served a double purpose. Since he wrote in vernacular English, he reached a large audience of natives and, more importantly, succeeded in having his theory ignored by Scholastics who, as purveyors of truth, would not stoop to read a vulgar throwaway journal. Censors would tend to downplay the relevance of such scribbles as well, particularly since the essay was devout enough and no one became upset. To that extent, *A Perfit Description* was a perfect deception, but the common person in England learned of Copernican heliocentricism through a diagram that incorporated the idea of an infinite Universe of stars as well, which gave rise to the mistaken impression that this concept was Copernican. Consequently, Digges' essay has not received the priority that it deserves, the more so because he wrote in English at a time when Latin was the language of erudition.

Eight years after Digges' essay, Bruno's *De l'Infinito Universo e Mondi* concluded on metaphysical grounds that the Universe contained an infinite plurality of worlds. In 1584, Duncan Liddell lectured on cosmology and made no mention of Digges' model. In 1600, William Gilbert's *De Magnete* asserted that stars are at varying distances and, although he looked favorably on the rotation of the Earth, he remained skeptical of heliocentricism and ignorant of the Digges model. In 1611, Donne prophesied the condemnation of Copernicus and Galileo in the afterlife, but he made no prediction for Thomas Digges.

In 1620 or later, Robert Burton wrote of "Heavens ... above the Firmament" which have "lately [been] revived by Copernicus, Brunus, and some others." By attributing the Diggesian model to Copernicus and Bruno, or including it incidentally in the catchall phrase "some others," Burton indicates that hardly anyone recognized the Diggesian contribution at that relatively late date. In 1841, Halliwell wrote that Thomas Digges "made no great contribution to science." In 1898, Berry (sect. 95) dismisses *Alae seu* as "an astronomical treatise of no great importance" and does not mention his other works. Kocher (218n11) writes that the Copernican model embraces an infinite Universe and Olson, Olson and Doescher (71-3) call Digges' model a "universe according to the Copernican heliocentric system." These examples illustrate what Johnson and Larkey (71-4) have noted, that, historically, even when Digges receives mention, the citation is incidental to the Copernican achievement.

In addition, historians did not check the publishing history of *A Prognostication Everlasting*, so the idea gained currency that Digges made his ideas known only in the 1592 edition, eight years after Bruno's ideas became known. Once established, the damage done by the "disregard syndrome" (Ginsburg; Summers) is hard to undo. The Diggeses were so successful in staying out of the limelight that, to this day, scholars ignore or trivialize his accomplishments, even in astronomy textbooks (Best *et al.*; Maene *et al.*).

Paradox.

At a stretch, a feature that could amount to a bona fide criticism of the Diggesian hypothesis is the homogeneity of the stellar distribution implied by the regular spacing of stars depicted in Figure 6. Digges fails to address what Edmond Halley (1656-1742) noted in 1720, that an infinite number of stars of finite size should light up the sky and make it as bright as the surface of any one of them, which it plainly is not. This is the de Cheseaux-Olbers Paradox, named for Philippe Loys de Cheseaux (1718-1751) and Heinrich Olbers (1758-1840). In 1848, in *Eureka*, Edgar Allen Poe (1809-1849) first successfully addressed the problem. By then Olaus Roemer (1644-1710) had demonstrated the finiteness of the speed of light and Poe conjectured, correctly, that the Universe is too young for all the light from distant stars to have reached us. In Digges' day, no one had joined the issue of the speed of light in any meaningful way and, even if Digges had pondered this issue, he would not have had the scientific basis to address it.

Diggesian Precession.

Ptolemaic astronomy explained the phenomenon of Precession by keeping the Earth fixed and moving the entire sphere of stars. A moment's reflection reveals, however, that the two geometrical entities necessary to define Precession, viz. the Celestial Equator and the Ecliptic, arise precisely from the two new Copernican motions, the rotation and revolution of the Earth. Copernicus did not have a firm grasp of the concept of inertia, but the net result of his confused arguments was that the sphere of stars is fixed, that the Earth rotated and revolved and that it also wobbled. In effect, he removed the onus of the phenomenon of Precession from the sphere of the stars and placed it on the Earth and, thereby, corrected the old mistake of attributing to appearances what in reality belongs to the observer.

Digges' stars extend indefinitely outward and, as a whole, define the standard of rest because it is absurd to think that stars scattered at arbitrarily great distances can spin around the Earth daily and in perfect unison. Digges agreed with Copernicus that the Earth, and not the distribution of stars, bears responsibility for Precession, but declined to address the matter further because Precession is not essential to his case and plays a relatively minor role in his grand vision. Digges set priorities by identifying the parts of the problem relevant to his larger purpose. Science is progressive, and Digges did not solve all problems at once any more than Rome was built in a day.

Everywhere or Nowhere.

Nicholas of Cusa held that no object could lie at the center of physically infinite space because the center would have to be everywhere, or nowhere. England's best mathematicians would know this, yet Kuhn (*Copernican* 233-4) says that, in positing an infinite Universe, Digges makes the conceptual error of supposing that the Sun and Solar System lie at the center of an infinite distribution of stars. Examination of Digges' text, however, reveals that this is not what he says. What Digges says is that "we can neuer sufficiently admire thys wonderfull & incomprehensible huge frame of goddes woorke ... [e]specially of that fixed Orbe garnished with lightes innumerable and reaching vp in

Sphaericall altitude without ende." This is a statement of appearances and not of reality. He says that the Universe of stars reaches outward (up in spherical altitude) without end, and that the observed "orbe" of stars is admirable in the minds of mortals. This does not imply that observers are at the center of anything other than of what they admire, and they admire what they can see. The object of admiration is a "fixed Orbe garnished with lights innumerable," which he defines as follows:

> We onely beholde sutch as are in the inferioure
> partes of the same Orbe, and as they are hygher,
> so seeme they of lesse and lesser quantity, euen
> tyll our sighte beinge not able farder to reach or
> conceyue, the greatest part rest by reason of their
> wonderfull distance vnto vs.

In other words, the stars of Digges' "fixed Orb" are those that we behold owing to their proximity, whereas less bright ones are invisible owing to distance. Digges explains that the limited sensitivity of the eye causes observers to see themselves at the center of a distribution of stars that reaches only to a certain distance but which, in reality, stretches far beyond the capacity of their eyes to descry. Thus, the supposed center wherein we live is apparent and not real because the Diggesian observer, who finds the huge frame of God's work as admirable as it is incomprehensible, is of necessity at the center of that which he admires and finds incomprehensible.

Many question the finitude of the Universe because a bound is impossible to visualize without something beyond binding it. In *Tractatus* of 1921, Ludwig Wittgenstein (1889-1951) noted, that in order to draw a limit to thinking, we must think both sides of the limit. For example, Lucretius could not conceive of throwing a javelin out of a finite Universe any more than a modern film star can rid the World of a soft-drink bottle by tossing it off the cliff of physical space. For present purposes, it suffices to say that Digges sees the huge frame of God's work as incomprehensible because, not only is an infinite Universe inherently beyond his comprehension, but also because the outer space of stars "may wel be thought of vs to be the gloriouse court of ye great god." The difficulty, it seems, is that no mortal can fully comprehend the Almighty, let alone infinite space at the same time. Thomas Digges invites human inquirers to think of cosmic space in terms that are both natural and supernatural, and physical and metaphysical, and to conduct the quest for knowledge accordingly.

To McLean (150), "Digges' conviction of 'stars innumerable' indicates some kind of optical penetration of space." Johnson and Larkey (105) write, "it becomes of the utmost importance, in evaluating [Thomas] Digges' work, to determine whether his idea that the Universe was infinite resulted solely from metaphysical considerations, or whether scientific reasoning and observation played the decisive part in the formulation of his theory." Before examining Thomas Digges' work in more detail, however, it is best to place it in the context of his father's invention of the perspective glass, on the off-chance that the New Philosophy achieved a major impetus from Leonard's dabbling in the natural magic of optics.

CHAPTER 4: OPTICS AND THE NEW ASTRONOMERS

Many evils betide him that revealeth secretes.
Roger Bacon

Lenses and their properties have a long history dating at least as far back as the Golden Age of Muslim culture. Ibn al Haitham (965-1039), known as Alhazen, wrote *Kitab al-manazir* on optics in which he attributed the origin of light rays to the object in view and not the eye, and theorized that the curvature of lenses causes magnification. He knew that the vertical displacement of celestial images near the horizon was owing to refraction of light by a finite atmosphere, which he assumed accurately, was about 10 miles. He was also the first to mention the camera obscura and to construct parabolic mirrors. In England, the first evidence for the study of lenses and optics comes from reports by Robert Grosseteste who taught at Oxford in the thirteenth century. In a treatise on rainbows, Grosseteste stated that lenses show how to "make things a very long distance off appear as if placed very close, and large near things appear very small." As a result, observers can "read the smallest letters at incredible distances." At about the same time, lens grinding became an important industrial activity at the leading center of glass making in Venice, Italy.

Roger Bacon.

Roger Bacon built a reputation as impressive as that of his predecessor, Grosseteste. Bacon's erudition extended to Latin, Greek, Hebrew, philosophy, medicine, law and divinity, and he was a staunch advocate of the experimental method. He conflicted with ecclesiastical authority and spent time in prison. He was saddled with a charge of sorcery, which he went to great lengths to refute, without success. From the time of his death, his reputation was "covered in layer after layer of myth and confusion" (Clegg 3). In 1556, Recorde wrote that the common man attributed Bacon's results to necromancy and evil spirits, but he defended him by stating that Bacon had no knowledge and expertise in the occult arts and attributed optical phenomena to natural causes. Denigration of Bacon's accomplishments persisted into the twentieth century, as, for example, when historians labeled him an armchair scientist who made no original contributions to knowledge. It is true that Bacon conjured up visions of horseless carriages and flying machines, but these were the product of his fertile imagination and, in any case, were no more outrageous than the contraptions envisioned by Leonardo da Vinci (1452-1519).

Itinerant peddlers sold Venetian glass artifacts throughout northern Europe, some of which might have found their way to England. If Grosseteste and Bacon bought glass items for experimental purposes from commercial sources, they would not have been the last to do so. In the seventeenth century, Galileo procured his first lenses from an optical shop and Newton bought a glass prism at Sturbridge Fair in order to study the phenomenon of color. Through the thirteenth century in England, however, glass making made considerable progress and Grosseteste and Bacon may have made their own instead of being mere passive consumers of artifacts procured from the merchants of Venice.

Bacon discovered the law of reflection as well as an approximation to the law of refraction, both in the context of glass lenses and their ability to magnify images. In a brief passage from Bacon's *Discovery of the miracles of art, nature, and magick* of about 1250, we learn of, "*glasses so cast*, that things at hand may appear at a distance, and things at a distance ... at hand ... letters may be read, and ... starres shine in what place you please" (emphasis added). The emphasis draws attention to two meanings for the verb "to cast," both extant from about the end of the thirteenth century (*OED*). The verb falls into the general category of "to put into order" and "to arrange" and within that category, "to order" or "to lay out in order" a piece of work or other material thing, or to form into a shape by pouring a soft or molten substance into a mould. Thus, "casting" could mean either "shaping" or "arranging." [I am indebted to van Helden ("Invention") for some of the quotations in this chapter.]

In *Opus Majus* of about 1267, Bacon describes in detail the results of his experiments on reflection and refraction using lenses and mirrors and concludes his long and detailed account with a short chapter, one paragraph long, in which he clarifies the distinction between "shaping" and "arranging." He writes (emphasis added), "For we can so *shape* transparent bodies, and *arrange* them in such a way with respect to our sight and objects of vision, that the rays will be refracted and bent in any direction we desire, and under any angle we wish we shall see the object near or at a distance. Thus from an incredible distance we may read the smallest letters." The first sentence of this passage is a compound one. From its first part, we understand that Bacon *arranged shaped* glass lenses ("transparent bodies") along the line of sight to various "objects of vision" in such a way that he could refract the rays from them in any way he pleases. In the second part, he writes of "the object," which restricts attention to a particular object. Although Bacon does not say whether he views that single object with one or more lenses, the fact that he speaks of arranging a plural number of lenses in the same breath as he refers to a single object of interest, suggests that he views a particular object with more than one lens at a time.

Bacon systematically examined reflection and refraction by both convex and concave spherical surfaces and, for as thorough an experimenter as he was, it seems reasonable to suppose that he investigated the effects of pairing lenses using different "casts." If so, he would probably have discovered the need to separate the lenses by more than an arm's length because lenses with relatively short focal length were difficult to make. Regardless, he would find it hard to get results by holding lenses in his tremulous hands and might have thought of securing them in desired positions and orientations using simple structures of wood or metal. In this way, he could position lenses in a continuous variety of ways and investigate properties in a systematic and thorough manner. It is not beyond reason to wonder whether he discovered the so-called Keplerian or "telescopic" combination in which rays from the convex objective (front) lens pass through the focal plane and a second convex lens converts the diverging rays to ones that the eye can bring to a focus.

Concave lenses made their official debut in England only in about 1450 (van Helden "Invention" 10). However, Bacon had examined their properties more than a century earlier. It seems reasonable to suggest that only a less-than-thorough series of experiments would fail to turn up the lens combination of a convex and a concave lens, the so-called Galilean or "astronomical" cast that emerged in Holland in the early seventeenth century. Unfortunately, Bacon's brief account supplies little detail, which fits a style in which scientists develop a general theory and end their dissertation with a few simple examples. This would explain the cryptic commentary that, with optical aid, "a small army might appear very large, and situated at a distance might appear close at hand, and the reverse [and that] the sun, moon, and stars in appearance ... descend here below." The evidence suggests that Bacon discovered a two-lens telescope, and Clegg (48-9) agrees. Bacon knew that the Milky Way is comprised of vast numbers of stars too close together for the naked eye to distinguish, suggesting that he examined *Galaxia* telescopically.

His capacity for abstraction is evident when he theorized that the material of the bounding eighth sphere of stars must be transparent to outgoing radiation because, otherwise, it would trap light from the Sun and stars and the Universe would become steadily brighter. This argument places yet another demand on the nature of the bounding sphere, that it act as a one-way mirror by passing light outward but allow none to enter from the brilliant Empyrean.

Optics after Bacon.

Bernard de Gordon (d. 1314) and Guy de Chauliac (1298-1368) debated whether to correct vision with eye lotions or eyeglasses and, in the 1430s, some paintings actually showed mirrors and lenses. One by Robert Campin (1375-1444) shows a small convex mirror that reflects rather well an image of a whole room, and another by Jan van Eyck (*c.*1390-1441) shows a Church official holding a pair of spectacles. In the late 1420s or early 1430s, painters suddenly switched to greater naturalism, suggesting the use of optical imagery. The new "optical look" depicted curved surfaces accurately, as in *The Ambassadors* by Hans Holbein (*c.*1460-1524), which shows in realistic perspective the scales and coordinate lines of scientific instruments and globes. By the early sixteenth century, Leonardo da Vinci had sketched lens-grinding equipment in his notebooks. In 1538, Girolamo Fracastoro, who was a physician and a prodigious writer on astronomy, stated that two spectacle lenses "one placed on top of the other" magnify the Moon or "another star," i.e., Ancient Planet. Fracastoro probably had the makings of the Galilean spyglass because, in his time, physicians had both convex and concave lenses at their disposal. Some dispute these early claims, yet Fracastoro's description is virtually repeated by Giovanbaptista della Porta (1535-1615) who wrote in 1589 that "if you know how to fit [convex and concave] lenses together, you shall see both things afar off, and things near at hand, both greater and clearly." Taken at face value, Porta's words indicate that he had assembled a Galilean spyglass, although later, in 1609, for no apparent reason, he devalued his invention by calling it a "hoax." In those days, many in authority believed that optical imagery was fraudulent, so perhaps Porta took a hint and repented.

The Galilean design had been anticipated also by Raffael Gualterotti (1548-1639), who constructed a spyglass in 1598 for military use and not, as he states explicitly in a letter to Galileo, for the study of the stars. A combination of lenses was in use on the Continent at least as early as the turn of the seventeenth century and, through the first decade of the seventeenth century, the fledgling two-lens design continued to attract attention from professional soldiers. In 1609, the concept reached Galileo, who immediately set about making a spyglass. There was, however, an earlier contender.

The Perspective Glass.

In 1571, Thomas Digges announced his father's death in *Pantometria*. In it, he describes his father's labors of love. Leonard, "by his continual painfull practises, assisted with demonstrations *Mathematicall*, was able, and sundrie times hath by proportional Glasses duely situate in convenient angles, not onely discovered things farre off, read letters, numbered peeces of money with the very coyne and superscription thereof, cast by some of his freends ... in open fields ... but also seven myles of[f] declared wat hath beene doon at that instante in private places." Thomas Digges mentions accomplishing "many other matters farre more straunge and rare," but does not elaborate.

In 1579, Thomas Digges published *An Arithmetricall Militare Treatise, named Stratioticos* dealing with the science of numbers and military matters. He relates how his father's interest in optics "grew by the aide he had by one old written booke of the same *Bakons Experiments*, that by straunge adventure, or rather *Destinie*, came to his hands." Thomas hastens to add that Bacon's book alone did not do the trick. He says that his father acquired knowledge chiefly by conjoining continual laborious practice with mathematical skill. By 1579, therefore, the Diggeses had made advances in mathematics, military science, and optics. Thomas reports that his father could detonate explosives from a distance of half a mile, presumably with sunlight and mirrors. The emphasis on mathematics testifies to Leonard's combination of theory and practice and, in particular, suggests his familiarity with paraboloidal mirrors whose unique features Alhazen had studied several centuries earlier.

Leonard Digges' telescopic magnifier comprises a convex or plano-convex objective lens and a spherical mirror that serves as an eyepiece, the so-called "mirror-lens." The convex lens brings light rays to a focus, beyond which the rays diverge, and the mirror-lens then converts these diverging rays into ones that the eye can bring to a focus. The center of curvature of the spherical mirror is displaced slightly from the optical axis of the objective lens so that that the spherical segment lying on the optical axis can reflect light at an angle to the axis, which the eye can then access (see Rienitz *Historisch* 106-110).

The perspective glass takes its name from the Latin *perspicere,* meaning "to look through" or "to see clearly." In Latin, the device is a *perspicillum*, which is not to be confused with so-called "perspective" drawings of the fifteenth century. Allan Mills notes that the perspective glass has much in common with the Schmidt telescope, both of which have certain features in common with the eye of a sea scallop (Darius 10).

William Bourne.

A 1578 book by the Gravesend shipwright William Bourne (*fl.*1565-1588) contains much of what we know about perspective glasses. In *Inventions or Devices. Very necessary for all generalls and captaines, or leaders of men, as well as by sea as by land,* Bourne describes the capabilities of the perspective glass. He reports that a lens-mirror combination magnifies distant objects and describes the second optical element, the mirror-lens with its polished side facing the lens. He writes of convex lenses up to 16 inches in diameter and a quarter of an inch thick in the middle. He describes what one sees through them, but pointedly omits details. In effect, all he says is that "you shall see a small thing [at] a great distance ... this is very necessary ... as the viewing of an army ... which I doo omit." Bourne had military uses in mind and so did John Dee who advocated placing a perspective glasses in every one of Her Majesty's ships.

In about 1585, at the urging of Lord Burghley, Bourne composed another work entitled *A treatise on the properties and qualities of glasses for optical purposes, according to the making, polishing, and grinding of them.* In this, he tells of convex lenses made of "fyne and white Vennys Glasse" ground smooth by a concave tool of iron. The lenses are thin, have long focal lengths, and are set in frames to help prevent fracture. Bourne makes no direct mention of the quality of magnified images produced by a single concave lens, which, for many people, would appear fuzzy. Instead, he says that an image (emphasis added), "*especially* ... will be much amplifyed and *furdered*, by the receavinge of the beame that cometh thorow the glasse, somewhatt concave or hollowe inwards and well polysshed." From the time of Alfred the Great (849-899?), the verb "to furder" means to further, assist, promote, or favor an action (*OED*). Therefore, it appears that a second element *especially favors* the amplified image, which may mean that it improves its quality. Bourne says that concave mirrors are "very necessary for perspective," and he mentions the sizes of lens and mirrors. A second element, the mirror-lens eyepiece, is, therefore, an indispensable element of the design.

Bourne goes on to describe metal convex and concave mirrors that are well polished and well foiled. Only in the ninth and final chapter of this second work, however, does he get around to describing convex lenses and concave mirrors in combination. He reaffirms that Leonard Digges built a perspective glass "withowte any dowbte of the matter," but details are sparse and he defers to Dee and Thomas Digges because, he says, they know more than he does. Bourne blames poverty, inexperience, and lack of time for the sudden dearth of detail, conditions that Burghley, surely, could have alleviated if he had had a mind to. Bourne's treatise establishes a precedent for the design and general functioning of the perspective glass but divulges little that is new, and its occasional inconsistencies suggest that he was unfamiliar with some aspects of what he was describing. If Burghley had wanted a thoroughgoing exposition, why did he not ask Dee or Digges to write one?

Thin lenses promote transparency but have long focal lengths. A convex lens one foot in diameter and one-quarter inch thick has a focal length of several yards, which, in combination with short focal length mirror-lenses, increase magnification and tend to reduce the effective field-of-view. The image quality of the lens-mirror also degrades with increasing angular distance from the optical axis owing to spherical aberration, a phenomenon known and studied centuries earlier by Alhazen. Other optical aberrations were almost certainly present, further limiting the usable field-of-view. Without knowing details of the optics, we cannot know the actual capabilities of the device or the size of its usable field and must rely on Bourne's fragmentary comment that "the greatest impediment ys, that yow can not beholde, and see, but the smaller quantity [field-of-view] at a time." Experts believe that this usable field is a good deal less than the approximately 15 minutes of arc available to Galileo.

The resolving power of a telescope is its ability to discern detail in an image. Other things being equal, resolving power improves with the increasing aperture of the objective lens, but only up to a point, because atmospheric scintillation, the "seeing," sets the limit of resolution for objective lenses larger than a few inches. Bourne describes the resolving power of a perspective glass by saying that it discerns script from a distance of a quarter-mile. For example, symbols that are two-fifths of an inch in size, seen from a distance of a quarter-mile, subtend an angle of about five arc seconds. To see letters distinctly, the telescopic resolution of the perspective glass has to be less than this, say about one arc second or so. In theory, this is achievable by an aperture with a nominal size of a few inches or more.

Bourne speaks of foil-backed mirrors, but the Diggeses may not have needed them because, in the sixteenth century, reflecting glass coated with metal amalgam was in general use. If Leonard Digges could make a paraboloidal mirror, he could have made the elements of the perspective glass, particularly if he had the blessing of the government.

Modern ahistoric replicas of the perspective glass exist, constructed from the Bourne descriptions. The first known is by Joachim Rienitz in Germany. Others are by Colin Ronan, Allan Mills, Gilbert Satterthwaite and Howard Dawes in Britain, and Ewan Whittaker in the United States. However, current opinion remains divided on whether an Elizabethan telescope existed. Howse sums up the division, "There is no positive evidence that there was an Elizabethan telescope; but equally, there is no evidence that there was not." Improbability is insufficient to preclude possibility, just as, to cite a well-known instance, the improbability of the Earth's motion does not preclude the possibility of it.

Dreams.

Perspective glasses of good quality and large light-gathering power would be hard to make and would present enormous difficulties with rigidity, target acquisition, tracking and focusing. For these reasons among many, historians have doubted Bourne's claims. This lack of information provides cover for skeptics who dismiss early accounts as dreams, fantasy, fiction, speculation, embellishment, extravagance and exaggeration, and attribute tales of optical magnification to Magi who manipulate nature in order to deceive the senses.

On the other hand, Rienitz ("Glasses" 7) writes that it is "inconsiderate" to dismiss early accounts of optical feats as fanciful, "for in the old days some people were concerned with very substantial things too." Fred Watson (51-3) wonders whether in the sixteenth century "someone, somewhere, might just have succeeded in making a device to see distant objects," and advocates leaving the possibility open pending scientifically verifiable evidence like an intact relic or an unambiguous drawing with incontrovertible provenance.

Recorde expressed the view that princes should know of optical matters but not commoners. It is hardly credible, therefore, that England's foremost mathematicians lived in a dream world in which telescopic devices were described but not built or used. Many doubt that the Diggeses trained the device on the heavens, not even to peer at the Moon, but surely it must have occurred to them that stars are useful sources of dim, collimated light. It is hard to see why Leonard Digges, the inventor of the theodolyte, would write a book of fiction devoted to surveying and why Burghley, Bourne Dee and Digges would go to all the trouble of promoting a device that did not work or did not exist. In light of military threats to the island nation, it is unlikely that Bourne and associates were trying to scare enemies with tales of English sorcerers especially since, toward the end of the sixteenth century, potential enemies on the Continent had begun to toy with spyglasses.

Thomas Digges was a super-patriot who practiced self-censorship to a fault and who, for seemingly contrived reasons, ceased publishing at a critical juncture in his career. In *Stratioticos*, Thomas Digges lists books that he had begun writing and that he intended to publish, but never did. The list includes works on navigation, architecture, artillery, pyrotechnics and fortification. Especially germane to the topic at hand is an unpublished commentary on the revolutions of Copernicus that, he says, ratify and confirm his theory by evident demonstrations grounded upon late observations. We must wonder what these were. This work would have elaborated on *A Perfit Description*, which, retrospectively, seems merely a stopgap measure in the overall course of inquiry.

Digges says that he would have completed these works except that "the Infernall Furies" so tormented him with "Lawe-Brabbles" that he had to discontinue his studies. Anthony à Wood does not buy the excuse, saying that distractions other than law-brabbles contributed to the cessation of scholarly output. Had the promised work appeared, it would have put an end to speculation and have obviated the need for the present comical-tragical-historical-scientifical entertainment.

At the end of the Preface to *Stratioticos*, Thomas Digges refers to Pythagorean "exclusiveness," by which he may mean that he transmitted wisdom by word of mouth to an initiated few. Thomas Digges tries to justify his reluctance to publish by referring idiomatically to the precedence set by his father as *per manus tradere,* i.e., the handing down of knowledge from father to son, by which knowledge is committed to memory and propagates orally. Perhaps Leonard Digges admired Pythagoras and imitated him by divulging secrets of optics to his son who in turn transmitted them to a few select friends.

Perspective, Poesy, Politics.

This was the time of the institutionalization of Protestantism in England and of threats from Catholic countries. Factions loyal to Mary, Queen of Scots (1542-87), and her French allies menaced England from the north. In 1555, Philip II of Spain left England after failing to be crowned king and, after the death of Queen Mary, tried again by proposing marriage to Queen Elizabeth. Elizabeth turned him down and Spain emerged as England's chief adversary. In retaliation for raids on Spanish shipping, Philip vowed to invade England, and he assembled a fleet of warships to that end. After the Spanish Armada was defeated in 1588, Digges warned of a new threat of invasion from France. He drew attention to forces assembled at Calais and Dunkirk that were a potential menace because, among several reasons, the troops were idle and had nothing better to do.

Given the atmosphere of religious intolerance, it would not do for the world to know that English mathematicians were heretical and that, if anyone just happened to believe that the images they saw in the sky were not figments, that they had the means to see them. Just as the functions of spy satellites are not common knowledge in modern times, so the English would favor keeping quiet about their optical perspectives. Two connected facts emerge: Burghley was a lover of learning and a patron of the arts who had an abiding interest in cosmic phenomena, and the Diggeses pursued research of interest to the military-scientific complex. The commonality of interest is apparent from Thomas Digges' dedication of *Alae seu* to Burghley.

No one trifled with this spymaster, least of all innovative thinkers who relied on him for protection. Under *Regnum Cecilianum*, those who stepped out of line received harsh treatment. In England, the period 1580 to 1603 was the heyday of the practice of torture, which was not limited to mere physical pain but included psychological strategies designed to induce terror. Burghley exerted power through a system of patronage, thereby creating an extensive network of obligation to which the Digges family likely belonged. Moreover, Elizabeth was a no-nonsense queen who did not suffer fools gladly and would not have tolerated jugglers who spun fairy tales, let alone have her chief minister commission reports from one of them. Dee demonstrated the perspective glass to the queen to her delight and it is simply not credible that she, Burghley and the mathematicians suffered collectively from an optical delusional disorder.

At the same time, members of the Privy Council were concerned with politics and probably had next to no interest in the stars and planets. Thus, Thomas Digges might have felt free to regard information on celestial objects as grist to the mill of artistic creativity and to have relayed data to a capable poet *per manus tradere.* At the very least, if not actually involved in national security, the poet would need to be a member of the nobility and one of rare wit. Edward de Vere would be such a person.

By 1596, at least one poet had noticed the existence of optical glasses. In *The Faerie Queen,* Edmund Spenser describes a feat of the legendary magician Merlin:

> The great magitien Marlin had deviz'd,
> A looking-glasse, right wondrously aguiz'd,
> Whose vertue had to shew in perfect sight
> Whatever thing was in the world contaynd
> Betwixt the lowest earth and hevens hight
> So that it to a looker appertaynd:
> Whatever foe had wrought, or frend had faynd.

Spenser alludes to the common perception that optical imagery is magical, but points out the military and domestic applications of the spyglass and refers specifically to the upper limit of its utility as "hevens hight." In *All's Well that Ends Well*, Bertram speaks of the phenomenon of "perspective" which warps lines and extends or contracts proportions, i.e., of an optical glass that distorts and magnifies images (Bevington *All's Well that Ends Well* 5.3.49n). In *Twelfth Night*, Orsino speaks of "a natural perspective," which could be "an optical device" or "an illusion created by nature" (Bevington *Twelfth Night* 5.1.216n). In *Richard II*, Bushy says:

> For Sorrow's eyes, glazed with blinding tears,
> Divides one thing entire to many objects,
> Like perspectives, which, rightly gazed upon,
> Show nothing but confusion; eyed awry,
> Distinguish form.

The lines may refer to "trick paintings" that appear grossly elongated when viewed face-on but shrink into a recognizable image when viewed at an acute angle. A well-known example appears in Holbein's *The Ambassadors,* which features an anamorphic skull that he copied, supposedly, from a distorted image that he saw in a curved mirror. Bushy's words are also a fair description of the operation of a perspective glass. When the mirror-lens reflects light at an angle to the optical axis so that the eye can receive it, the image is viewed "obliquely" or "awry" in order for the "perspective" to "distinguish form" (Bevington *Richard II* 2.2.18n). Bushy speaks of tearful eyes that blur images; i.e., "sorrow's eyes" that are glazed with blinding tears and divide "one thing entire to many objects." A segue to "perspectives" follows immediately. Properly used, perspectives "distinguish form." There is a similar metaphor in *Hamlet* in which teary eyes "make milch the burning eyes of heaven" (see Chapter 10). The intimation is that perspective glasses resolve images that are otherwise confused.

The comparatively early dates estimated for the writing of *The Faerie Queen* and *Richard II* suggest that, before the writing of *Hamlet*, poets were sufficiently well versed in the capabilities of perspective glasses to write of them. In *Hamlet* itself, Ophelia muses on what she perceives as Hamlet's loss of mind and refers to the "glass of fashion and the mould of form." This means (Edwards 3.1.147n) that people "shaped themselves ... after his [i.e., Hamlet's] pattern" because, from about 1330, "moulde" signified a person of distinction. The "glass of fashion" could refer to the mirror-lens used in perspective glasses that gives an "ideal image." "Mould" indicates the shaping or fashioning of something, the "something" in this case being the lens. In the next scene, Hamlet exhorts the players "to hold as 'twere the mirror up to nature," referring to the mirror "which sets standards ... by revealing things not as they seem, but as they really are" (Edwards 3.2.18n). The mirror-lens serves is an eyepiece and the "mould of form" describes a convex lens, which, together, comprise the two optical elements of the perspective glass.

Celestial observations occurred on the Continent as early as 1598 and Burghley and his innovators would realize that it was simply a matter of time before the English design was re-discovered. The putative English litterateur would hardly wish to help this process along, nor risk divulging state secrets. Men of letters had only one recourse: they must protect both scientific priority and the national interest by writing in such a way that the underlying meaning would not readily spring to mind, yet not so cryptically that the true story would never emerge. Encryption was common. For example, in 1610, Galileo composed an anagram to establish priority of discovery. In that case it was clear that there had been one, but Digges' confidant had to be more circumspect. Digges and his poet knew that it would be simply a matter of time before someone would re-discover facts about celestial objects. This meant that the poet had to increase the opacity of encryption in order that the meaning would be obscure. The actual discoverer and his scribe could remain under the radar while the luckless re-discoverer would take the heat. History records that that someone was Galileo Galilei.

Connections.

Tycho circulated copies of his *Liber Secundus* among friends and acquaintances and, in 1590, wrote to one of England's most learned men, Thomas Savile (d. 1593), enclosing two copies along with four copies of his portrait (see Figure 8). In his cover letter, Tycho asked to be remembered to John Dee and Thomas Digges, both of whom had written books on the New Star of 1572. In that same year, Tycho was the honored recipient of epigrams from the Royal Chancellor of Scotland and the Scottish monarch, James VI (1566-1625), and felt driven to seek approbation from English poets as well. Savile may have honored the astronomical side of Tycho's request by forwarding Tycho's portrait and book to Thomas Digges because a copy of Tycho's portrait ended up in the possession of Digges' son, Dudley, but, as far as we know, no one in England took up the challenge of Tycho's request for poetical acclaim. If a poet did accept, he would have to know astronomy quite well, and what better way to learn of it than to receive lessons from England's leading astronomer and mathematician, Thomas Digges.

Honigmann (*King John* 2.1.574n) believes that Hotson (*Appoint*) has proved William Shakspere's connection to Thomas Digges, a view corroborated by Rowse (*Man* 197, 225-6). Shakspere lived near the Digges' home in London, and after the death of Thomas Digges in 1595, Digges' widow married Thomas Russell (1570-1634), the overseer of Shakspere's will. In the parish of St. Mary Aldermanbury near Guildhall, William Shakspere was a frequent guest at the house of actor and playhouse manager John Heminges (1566-1630), which was near to the Digges home. The theory is that Shakspere came to know the

Diggeses there. After 1600, Thomas' second son, Leonard Digges the Younger (1588-1635), became friends with Shakspere in Stratford and was a natural choice to compose the encomia published in the First and Second Folios.

On the other hand, Whalen argues that a de Vere-Shakespeare-Digges connection is plausible. Halstead points out that John Dee taught de Vere astrology and that Dee, who had Thomas Digges as a pupil, may have alerted de Vere to the latest advances in astronomy. Thomas Digges and Edward de Vere were only about 4 years apart in age, and it is common for students of the same teacher to know one another. The wife of Edward de Vere's tutor, Sir Thomas Smith (1513-1577) was first cousin to the wife of Leonard Digges, and de Vere and Thomas Digges probably became acquainted in that way (Hughes "Thomas Smith"). While a teenager at Cecil House, Oxford would likely have met Barnaby Googe who visited frequently. Oxford's tutor Smith was a leading mathematician who was well-versed in matters cosmological and his library contained Ptolemy's *Almagest*, Copernicus' *de Revolutionibus*, and the works of Euclid (*fl.c.*300 BC), Peter Apian and the Prutenic Tables (Hughes "Thomas Smith"; Johnson *Thought* 89-90). Gabriel Harvey proclaimed Smith greater than Ptolemy and a hundred Alfonsos – Alfonso being the king who oversaw creation of the Alfonsine Tables; see Chapter 1. It seems plausible that Oxford had an abiding interest in astronomy and that Thomas Digges could have transmitted information to him before Digges death in 1595. Of de Vere's identified poems (May "Poems" 12-3) none contains more than a superficial reference to astronomy but, of course, he would have sought shelter in anonymity for reasons stated. For example, Thomas Watson's dedication of *Hekatompathia* to de Vere virtually identifies him as an anonymous contributor (Ogburn 661).

Hotson cites instances where Digges' works played a significant role in several Shakespearean plays. Shakespeare gathered his military information from *Stratioticos,* whose 1590 edition has passages closely resembling some in *Henry V.* Sohmer ("Crux") notes the influence of *Stratioticos* on *Othello.* Later we see that Shakespeare used Digges' *Pantometria* as another one of his sources, and it would strain credulity to believe that the Bard would ignore other Diggesian works.

Sophie and Erik.

The engraving that Tycho sent to England (see Figure 8) bears the names of his great-great-grandparents Erik Rosencrantz and Sophie Guildenstern. In 1910, Huizinga suggested that Shakespeare took these surnames from that engraving. He thinks this more probable than suggestions made in 1908 that Shakespeare latched onto these names from reports brought back by touring actors, or from the rosters of universities in Wittenberg or Padua where Danish students with these names enrolled, because these two names were common among Danish nobility.

If Tycho's portrait is the source, why did Shakespeare pick this particular pair when there were fourteen others in the picture to choose from? To forestall irrelevant speculation, Shakespeare would likely have selected names of a married couple, and settled on Rosencrantz and Guildenstern because these shields occupy favored places in the engraving. Both are on Tycho's right-hand side, with Sophie's closest to the ground and Erik's first on the arch directly above Sophie's.

Shakespeare entertains thought on multiple levels simultaneously and it is entirely possible that he has more than one source in mind. Tycho's third cousin Frederick Rosencrantz (1569-1602) and his "slightly less remote cousin" Knud Gyldenstierne (1575-1627) were both scholars at Wittenberg, both returned to Denmark in 1591, and went together to England in 1592 for about a year where they may have attracted the Bard's attention. These two ambassadors were virtually inseparable, prompting the epithet "Siamese twins" (Swank 12).

Supposition.

After Savile had received Tycho's letter and forwarded the pertinent information, it is likely that tension unfolded between the world's leading naked-eye astronomer and the world's only telescopic astronomer. At the epicenter, we surmise, lay the world's leading dramatist, a poet who was concerned not only with ideas that circulate in high places but with literature of all sorts, even a common almanac. Through contacts still in need of elucidation, Shakespeare saw through the Diggesian smokescreen and made a screen of his own that both documented and disguised the emergence of the New Philosophy. In the gravedigger scene in 5.1, in the midst of all the puns and double meanings, Hamlet says, "We must speak by the card or equivocation will undo us." Shakespeare must scatter an abundance of clues lest his disguise succeed too well and his message pass unnoticed forever; yet, of course, he must bury meaning sufficiently well to forestall the possibility of censure.

The perspective glass is a truly revolutionary device whose existence did not escape the attention of leading poets in England, but the absence of direct evidence prompts inquiry along different lines. Perhaps a connection exists between the astronomer whose optical instrument formed pictures of heavenly bodies and the Bard who painted pictures with words. If so, it behooves us to attend to the art of textual interpretation.

CHAPTER 5: INTERPRETATION

Oor universe is like an e'e
Turned in, man's benmaist hert to see,
And swamped in subjectivity.
But whether it can use its sicht
To bring what lies beyond to licht
The answer's still ayont my micht.

Hugh Macdiarmid: *The Great Wheel*

Interpretation is criticism aimed at uncovering meaning. The task is difficult because interpreters necessarily have limited knowledge and experience and see things from a unique perspective. Moreover, no one can know, retrospectively, the intention of an artist, so the best anyone can do is to lift the fog of uncertainty in a maximally reasonable way. The question of when interpretation is maximally reasonable requires the reader to face the dilemma that, for a work to have meaning, its individual parts must "hang together" and make sense as a whole, yet it is the overall interpretation that informs the meaning of the individual parts. This circularity poses a dilemma that appears to sentence readers to a state of perpetual ignorance, for if we cannot comprehend either the whole or the parts without the help of the other, it seems that no progress is possible. The quest for meaning is not a vicious circle, however.

Hermeneutics.

Hermeneutics is the study of the methodological principles of textual interpretation. Its fundamental canon is the so-called Hermeneutic Circle, or Circle of Understanding, in which information and explanation interact with one another dialectically. Armed with a collection of particulars, readers plunge in and posit a global interpretation, whereupon, oftentimes, they return to the particulars in order to improve their understanding of the whole.

The art of hermeneutics derives from the skills of Hermes, the fleet-footed messenger of the Greek gods, whose Roman counterpart is Mercury. In order to survive, Mercury developed a propensity for fast talk and faster footwork. His elusiveness matches that of his planetary namesake who cycles quickly in and out of view and is hard to spot in the twilit sky. Mercury's shifty character qualifies him to moderate the affairs of thieves, rogues, and scoundrels, and the affairs of science, at least as it was perceived in days of yore.

After the Reformation, scriptural interpretation no longer rested with a central authority and hermeneutics became essential to Biblical exegesis. More recently, it has become concerned with meaning and understanding of literature as a whole and of the individual texts that comprise that whole. The quest is like trying to understand literature's basic component, the sentence, which provides the context by which a reader divines the meaning of its words, except that, of course, the words themselves give meaning to the sentence. This sort of dilemma confounds attempts simultaneously to define a literary genre and to assign a text to it, since each depends on the other. Perhaps the Bard appreciated this difficulty when he makes fun of pedantic classifications of stage plays through Polonius' effete conflation, "tragical-comical-historical-pastoral."

Faced with a slew of puzzles, J. Dover Wilson (*Happens* 9, 321) feels it best to consider all simultaneously. Concerning *Hamlet*, he writes, "I came to see that the scientific thing to do was to attack all the problems at one and the same time, seeing that solutions must hang together, if *Hamlet* was an artistic unity at all." He foresees the need for an approach that is both global and particular, because to attack all the problems simultaneously is to step directly into the interpretive process inherent in the Circle of Understanding.

The Book of Nature.

Hermeneutics was once thought to be the sole province of the humanities but, in the latter half of the twentieth century, philosophers discovered that the metaphorical "book of nature" is subject to interpretation according to a similar set of rules. Centuries ago, Shakespeare had anticipated the discovery when he wrote, "In nature's infinite book of secrecy / A little I can read." Trying to understand Nature is like trying to understand a literary work except that, in science, the "texts" are the existents of the World and the World itself. All the while, of course, the "text" remains the final authority.

Examples from modern astronomy illustrate the iterative quality of scientific inquiry. Galaxy classification is not absolute (Sandage and Bedke) and certain stars that change in brightness are hard to pigeonhole (Zsoldos), yet preliminary typing is needed for research to proceed. Osterbrock describes one of many obstacles that astronomers must overcome in understanding quasars, "Ideally we should like to begin with a good working hypothesis as to the nature of an active nucleus, then on the basis of ... hypothesis calculate all the observational consequences, make measurements, and compare their results with predictions made on the basis of hypothesis. In practice of course one difficulty is that we do not have a clear physical hypothesis to begin with."

Because quasars are so bright, they are visible at cosmic distances of around ten thousand million light years or 100,000,000,000,000, 000,000,000 miles. Thus, quasars have the potential to serve as probes of the structure of the Universe. However, transmission of information occurs at a rate no greater than the speed of light, with the result that we see distant objects in the Universe not as they are now but as they were when the light was emitted. Cosmic space is expanding and curved, so light passing through it undergoes effects that are different than if it passed through space that is static and unaffected by the presence of matter and energy. Thus, to determine the structure and evolution of the Universe, it is necessary to observe objects that are now at enormous distances and whose physical attributes we do not completely know. In order to determine those attributes, we must make corrections for affects imposed on the light by the Universe, but of course, the structure of the Universe is what we set out to discover in the first place. Studies in cosmology require a sort of dialectic oscillation between its microscopic and macroscopic features each of which helps the other to hone a self-consistent solution (Stegmüller *Wissenschaft* 41; Usher "Hermeneutics" 182-3).

New Methods.

New ideas on scientific methodology played an important role in the overthrow of the Old Physics. The use of induction as a way to promote understanding is traceable to the scientific age of ancient Greece, when astronomers such as Hipparchus, Aristarchus and Eratosthenes incorporated it in attempts to understand the cosmos. In England, Robert Grosseteste understood the need for a new approach to understanding nature. Roger Bacon perpetuated the tradition and he, in turn, influenced the Diggeses.

In 1620, in *Novum Organum*, Francis Bacon (1561-1626) wrote of scientific methodology, but, as he was more of a philosopher than an experimentalist, he did not practice what he preached. However, in *Two New Sciences*, written earlier than its publication date of 1638, Galileo also advanced the new organon. The book contains a cogent statement of his experimental method of inquiry along with practically all that he had to say on the subject of physics. It is the cornerstone of modern physics and contains a succinct statement on what passes reasonably as the "scientific method." In a translation from 1730, Galileo writes (Berry sect. 134), "Let us therefore take this as a *Postulatum*, the truth whereof we shall afterwards find established, when we shall see other conclusions built upon this Hypothesis, to answer and most exactly to agree with experience." Galileo draws attention to the concept of a *postulatum*, or working hypothesis, that is the key to understanding in both the physical and literary sciences. Such a hypothesis may originate in that mysterious, even divinatory manner that leads to comprehending something with reference to what is known about other things. With a *postulatum* in hand, the inquirer takes an exegetic plunge into the Circle of Understanding, successive cycles of which lead, oftentimes, to interpretations that are ever more accurate and broadly based. This process is inductive to the extent that generalities follow from particulars.

Much is written of the methodological principles of scientific inquiry. To simplify matters, it suffices to assume here that beings-in-the-world are possessed of a sort of simian curiosity that compels them to gather and process information in order to make sense of the world around them. Thus, (1) observation and experimentation lead to the realization that there is something worth knowing or discovering, which (2) prompts the questioner to frame a hypothesis, (3) deduce consequences of the hypothesis in order (4) to test it. In simple terms, if the outcome of step (4) is positive, the inquirer returns to step (3) and continues to run tests by making further deductions or predictions. Call this loop, (4)-(3)-(4). If the outcome of step (4) is negative, then the inquirer returns to step (2) and modifies the hypothesis. Call this loop, (4)-(2)-(4). All the while, step (1) is ongoing so that the hypothesis (2) continues to benefit from fresh information. Steps (1)-(3) are re-entry points for the ongoing and essentially endless looping process whereby humans seek to determine physical reality from sensory input. Because hypotheses from step (2) and deductions from step (3) are of central importance in loops (4)-(2)-(4) and (4)-(3)-(4), Galileo's procedure is often called the hypothetico-deductive method. This closely resembles the hermeneutic-dialectic method of the humanities and, in both cases, doubt fuels the process (see Chapter 12).

The procedure of steps (1)-(4) can account for what Kuhn calls "puzzle solving" in "normal science" by which pieces of an existing theoretical structure or paradigm are tested and explained. When used in its most telling way, however, the predictions of step (3) are not of a "kind which is known," like making predictions from a known set of rules according to an existing paradigm, but are of a kind that demand a major revision or replacement of it. Such so-called "paradigm shifts" have occurred many times in the scientific age, the classic example being when heliocentricism supplanted geocentricism (Kuhn *Structure* 10-135; Popper *Conjectures* 117). When different hypotheses explain the same evidence equally well, Giles of Rome (*c.*1247-1316) and William of Occam offered a solution by which to favor one over another. The principle known as Occam's razor cuts out the theory with the most assumptions.

Objectivity.

Galileo receives credit for the formulation of the hypothetico-deductive method, but Digges' *A Perfit Description* has an earlier description. He states the attributes as follows: "There is no doubte but of a true grounde truer effects may be produced then of principles that are false, and of true principles falshod or absurditie cannot be inferred. ... If therefore the Earth be situate immoueable in the Center of the worlde, why find we not Theoretikes vppon that grounde to produce effects as true and certain as these of Copernicus?" In other words, a correct theory produces true effects or outcomes and a false theory does not. This recognizes the iterative nature of inquiry because it describes loops (4)-(3)-(4) and (4)-(2)-(4) according to which a hypothesis as mooted in step (2), will have consequences as deduced in step (3), which, when tested as in step (4), either support or deny it according to whether deductions are verified or not.

Hotson (*Appoint* 114) confirms that Digges favors "progressively" testing theories by observations and experiments over "the method of brilliant metaphysical speculation," which refers to Bruno who proposed an infinite Universe on philosophical grounds but, as a philosopher rather than a scientist, "never deigned to practice" the experimental method. By contrast, "Digges had the habit of testing all [scientific] ideas by actual experiments [as] proved by his ... books on scientific subjects" (Johnson and Larkey 99, 114).

Images.

Cosmic model-builders have as their goal the creation of a model of reality derived from cosmic appearances, a process that Pannekoek (102) describes as distinguishing between empirically detected phenomena and their true physical nature. For instance, a rainbow appears as a multicolored arc across the sky that emerges, we imagine, from a pot of gold, but is really an image created through reflection and refraction of sunlight by water droplets in air.

Copernicus distinguishes physical appearance and reality and illustrates the difference by relating an experience of Aeneas as recounted by Virgil, "We sail out of the harbor, and the land and the cities move away." Landlubbers see ships sailing away, but seamen see the land moving away. Just so, an observer at rest in the reference frame of the Earth thinks that the Sun and stars move, but Copernicus asserts that the Sun and stars are at rest and the Earth moves. Digges states that, "the true Motion in deede [is] in the Earth, and the apparance only in the Heaven." Senses untutored by the lessons of reality are easily misled. "Ptolemaic astronomy was Everyman's astronomy mathematised [whereas] Copernican astronomy bore not the faintest resemblance to everyday, common-sense" (Hall 459). Humans saw that the sphere of stars rotates and not terra firma and the idea that the Earth could spin seemed as preposterous as the Earth seemed ponderous. It was even more outlandish to think that the Earth could move through space. The image of an inverted bowl overhead is not what it seems either because, like the arc of a rainbow, the bowl of the sky is centered on the eye of every beholder. In a nutshell, the challenge is to take the "I" out of imagery.

Subjectivity.

In *To a Louse, On Seeing one on a Lady's Bonnet at Church*, Robbie Burns (1759-1796) illustrates the limitations imposed by subjectivity:

> O wad some Pow'r the giftie gie us
> To see oursels as others see us!
> It wad frae monie a blunder free us
> An' foolish notion.

Just as the louse and the churchgoer do not see eye to eye, so every exegete reads a text or approaches a problem from a necessarily limited perspective. This raises a host of problems.

The "documentary fallacy" treats a work of fiction as though it were a record of historical fact, from which inferences about other supposed facts could be drawn (Jenkins 123). Those who explain the past by wrapping themselves in their own modern consciousness are in danger of committing the interpretive sin of "presentism" (Hughes "Donne" 50). Interpreters suffer from a "confirmation bias" when they count hits and not misses, which they either ignore or rationalize into oblivion. Saul Bellow (367) warns of the risk of being a "deep reader" who latches onto a particle of philosophy or religion to the exclusion of others. Hotson (*Hilliard* 13-4) says that one of the worst ways to tackle a problem is to rush in with such conviction that when we encounter a difficult piece of evidence we promptly distort it into something familiar in order to support a "darling conjecture." Shakespeare recognizes this kind of subjectivity when a Gentleman remarks that those who listen to the seemingly incoherent chatter of Ophelia, "botch" her words to fit their own thoughts; i.e. patch her words up into patterns conforming to their own ideas (Edwards 4.5.10n). "Self-projection" is another form of subjectivity. Schoenbaum (*Lives* 330) remarks that bardolatrous amateurs usually suffer from this ailment, as do adherents to particular persuasions who convert Shakespeare to their "own belief or infidelity." According to Samuel Taylor Coleridge (1772-1834), every man sees himself in Shakespeare's plays "without knowing that he does so" (quoted by Garber 17).

When bardolatry stymies objectivity, passionate devotion can engender a disproportionately passionate interpretation. On the other hand, an all-too-common occurrence in literary interpretation is *not* ascribing meanings that the author *does* intend. This poses a dilemma that is hermeneutical in its own right because interpretation is suspect when it is beholden to the expertise of modern readers; yet, without new insights, no progress can occur. For example, before 1616 when William Harvey formally announced the discovery of the circulation of the blood, the Bard had demonstrated knowledge of that fact in at least nine plays written prior to 1608 (Davis), but the fact that Frank M. Davis M.D. is a physician is no reason for him to be silent. Shakespeare, they say, is for everyone. Anyone is a potential interpreter of the Canon and, in the case of *Hamlet*, amateurs have contributed successfully to understanding (Edwards 36; Levi p. xviii).

Some Shakespearean plays are "problem plays" because existing interpretations do not work well and, for this, the author often gets the blame, yet what seems odd or obscure may conceal some necessary points that critics have failed to grasp, in which case the question arises as to whose problem is whose. Wilson (*Happens* 14-5) urges readers to explore many options before blaming the author, lest they sin against the canon that seeks maximally reasonable interpretation.

Darling Conjectures.

A historical perspective is useful. *Hamlet* is regarded as the most problematic and enigmatic play in the Canon and the one most in need of explanation. Despite many analyses, *Hamlet* has remained a mystery, making it a popular target for new interpretations because everyone, eventually, wants to take a crack at it (Gottschalk 2-3; Wilson *Happens* 321).

James Plumptre (1771-1832) was no exception. He proposed that *Hamlet* was a censure of Mary, Queen of Scots, and he supported his claim with parallels between the historical and textual record. Gertrude marries her husband's murderer in haste, while Mary married Bothwell even though barely three months had elapsed following the murder of Lord Darnley in 1567. The incident of Polonius being killed in the queen's chambers parallels the death of Mary's advisor, Rizzio, in her apartment. The tardiness of King James in avenging his father's death parallels Hamlet's delay in dispatching Claudius. The topography around Elsinore resembles that near Salisbury Crags and Holyrood Palace. A Rosencrantz took Bothwell prisoner after his escape to Denmark, and a Guildenstern witnessed his deathbed confession. Dr. Wotton went to Scotland to spy on James and deliver him to England, just as Rosencrantz and Guildenstern attempt to deliver Hamlet to England. Plumptre was so enamored of his ideas that "every possible suggestion seemed additional proof to him" (Gurr 22; Johnston 180-6).

Some regarded Plumptre's zealotry as obsessional and, for several decades, critics treated his ideas with indifference. Hermeneutic cycling resumed in 1860 when further associations were developed and existing ones refined. Gertrude was associated with Mary Stuart, Hamlet with James, and Claudius with Bothwell. Laertes became the Laird of Gowrie who had a father's murder to avenge. However, "an air of burlesque" developed when someone noticed that Laird sounds like Laertes (Johnston 180-5) and, eventually, the Plumptre hypothesis fell by the wayside.

If the various parts of an interpretation do not hang together to form a rational whole, then that interpretation will not "work" (Wimsatt and Beardsley 469). Plumptre's idea did not "work" but it belongs, nevertheless, to the overall interpretive process. For example, Plumptre's conjecture is useful to the extent that he based his interpretation on historical events, an approach supported by Winstanley (166) who concluded in 1926 that it is "absolutely certain that Shakespeare is using a large element of contemporary history in *Hamlet*." If contemporary royal intrigue is not the case, what other contemporary history might have sufficient import to warrant such a work of art?

Personification.

Perhaps Shakespeare wrote *Hamlet* to reflect the revolutions in World view that were occurring in the latter part of the sixteenth century. Personification is a common literary tool and we posit that the "hero whom Shakespeare loved above all other creatures of his brain" (Wilson *Happens* 44) personifies the cosmological model of Thomas Digges (see Figure 6, Table 1). The Ghost is the spirit of Leonard Digges whose memory and works guided and inspired his son. Rosencrantz and Guildenstern personify the upstart hybrid geo-heliocentric model of the Dane, Tycho Brahe (see Figure 5), and Claudius personifies the bounded geocentric model of Claudius Ptolemy (see Figure 2). The *Aha! Erlebnis* is that the false king takes his name from Claudius Ptolemy whose falsity lay in his methodology and cosmic modeling.

Table 1

MODEL PERSONIFICATIONS

ROLE	MODEL	BUILDER	NATIONALITY
Claudius	Bounded geocentric	Claudius Ptolemy	Greco-Roman
Rosencrantz Guildenstern	Bounded hybrid	Tycho Brahe	Danish
Marcellus	Pre-Diggesian star distribution	Marcellus Palingenius	Italian
Horatio	Pre-Diggesian heliocentric	Nicholas Copernicus	Polish
Hamlet	Diggesian Infinite Cosmos	Thomas Digges	English

Claudius.

Bullough suggests however, that Shakespeare named Claudius for the Roman Emperor, Claudius I (10 BC-54 AD), who ascended the throne in 43 AD and conquered Britain the same year (Guilfoyle 42). If so, we expect that the emperor and the king would have something in common.

Legend has it that the Praetorian Guard selected Claudius to succeed the tyrannical emperor of ancient Rome, Caligula (12-41 AD). Apparently, a soldier had spotted a pair of feet sticking out from under some curtains and on discovering Claudius, carted him off to an uncertain fate. Nobody knew who the next Emperor should be, so Claudius got the job. At the time of his ascension, Claudius was married to his third wife, Valeria Messalina (22-48 AD), by whom he sired a son, Britannicus (41?-55 AD). Messalina was a meddler, which annoyed Claudius, so he had her killed. He then entered into an incestuous marriage with his brother's daughter, Agrippina II (d. 59 AD), who had mothered Nero (37-68 AD) in a previous marriage. She was a tireless advocate for Nero and persuaded Claudius to adopt him as his own. Claudius died, possibly from over-indulgence or, some say, from toadstools dished up by Agrippina. Nero inherited the throne when not yet seventeen years old, but Agrippina disapproved of her son's licentiousness and threatened to support Britannicus' claim to the throne. Agrippina herself had a considerable reputation for malfeasance and tried to seduce Nero, but the army warned Nero that this would never do, so the ingrate had his mother killed.

Like the Roman Emperor, Shakespeare's Claudius is gluttonous, marries incestuously, regards his new wife's son as his own and ingests poison. Some even think there is an intimation of incest between Hamlet and Gertrude. Identification with the Roman Emperor is unlikely, however, because few other aspects fit. Claudius I succeeded a tyrant but Old Hamlet was not a tyrant, and the Ghost specifically instructs the prince not to judge or kill his mother but to "leave her to heaven." Hamlet, in turn, affirms his resolve not to emulate Nero in any way, "O heart, lose not thy nature; let not ever / The soul of Nero enter this firm bosom."

Instead, suppose that *Hamlet* is an allegorical account of the historical struggle for acceptance of the New Astronomy. Literal and figurative meanings join like the real and imaginary parts of a complex number to contribute to the meaning of the whole. King Claudius thinks his cause is holy because he is both a king and a dogmatic defender of Christianized Aristotelianism, but the advent of a supernatural being imperils Claudius and, thus, the Ptolemaic World view. *Hamlet*'s terrestrial tragedy is a tale of power, position, and revenge, but the fact that we learn of the murder of Old Hamlet through the intervention of a supernatural being injects a new interpretive challenge because we must understand metaphysical concerns through effects discernable on the stage of the real world.

Appearance and Reality.

The difference between appearance and reality is basic to astronomy and is "the dilemma most persistent in Shakespeare" (Hunter p. xl). Shakespeare "is exceedingly conscious of the thin line between seeming and being, appearance and reality, and falseness and truth" (Rowse *Man* 103). In both real and supersensible space, the battle in *Hamlet* is between the spirits of good and evil – a duality that dominates the script. As the play begins, the bad spirit has already taken the initiative and has seated its surrogate on the throne at Elsinore. The good spirit cannot allow evil to triumph, however. The terrestrial political and the celestial supernatural games commence from the very first line, at which time the supernatural role of Claudius is already in place. At the start, the good spirits are still marshalling their forces and the dual role of Hamlet is not yet apparent. Hamlet is upset at conditions in the natural world, which we take to be the death of his father and his mother's hasty remarriage but, in reality, his disgruntlement has another, less obvious source. Before he sees the Ghost, Hamlet wishes to leave Elsinore, where the geocentric school prevails, and return to Wittenberg, where the heliocentric school prevails, but the royal couple persuades him to remain at Elsinore, which suits the supernatural agenda. Then the Ghost confounds Hamlet's sorrow by revealing the crime against his father and demanding that Hamlet exact revenge upon the perpetrator, Claudius. The natural and supernatural story lines co-exist because revenge is an emotion that exists in the real world, but the agent who commands Hamlet to exact revenge emanates from the supernatural world. The interpretive challenge is to disentangle appearance and reality, which prompts examination of precedent in Shakespeare's sources and the means by which the characters he invented achieve their ends. Ensuing chapters subject the present postulate to the test of evidence.

CHAPTER 6: MADNESS AND METHOD

Scientists are always prime targets of elites dedicated to doctrine.
Melvin Kranzberg

The chief source for Shakespeare's *Hamlet* is *Historia Danica*, written by the Danish historian, Saxo Grammaticus (*fl.*1188-1201). Saxo chronicles the exploits of a legendary prince, Amleth, who effected a mental disorder as a means to survive the malice of his incestuous king. Shakespeare models Hamlet on Amleth and, in particular, contrasts Hamlet's feigned madness with the deceitful methods used by the king and his court.

Literary Sources.

The relevant part of *Historia Danica* begins when Amleth's father, Horvendile, King of the Jutes, duels with Koll, King of Norway. Prior to the duel, they had agreed that if either one is maimed then he should perish, since he could never live with the humiliation. Horvendile severs Koll's foot and true to their pact, Horvendile kills him. For good measure, Horvendile kills Koll's sister, too. Having achieved fame in combat, Horvendile returns home, only to die at the hands of his brother, Feng. "Incest" caps this "unnatural murder" when Feng marries his murdered brother's widow, Geruth. Amleth feels threatened by Feng, who is now both his uncle and his stepfather. He feigns madness and resorts to double-speak in order to protect himself, but Feng suspects him of cunning and hatches a plot to test his mettle by luring him into the presence of a "fair woman." Amleth, alerted to the scheme, plans accordingly. As luck would have it, the temptress had been Amleth's childhood companion and is still fond of him, so they make love. To keep Feng from learning of this outcome, Amleth swears the damsel to secrecy and describes their meeting equivocally.

In a second test, a spy watches as Amleth visits his mother in her chambers, but Amleth kills the spy and chastises his mother. Feng is convinced of Amleth's guile, so he sends him to Britain with two companions who carry a message ordering Amleth's execution. Amleth

93

turns the tables on Feng by altering the message that the guards carry. On receiving the altered message, the British monarch cautiously decides to await developments. Amleth demonstrates his worth by showing remarkable insight in perceiving, correctly, that something is rotten both in the comestibles of a royal banquet and in the royal lineage. When the king discovers that Amleth is right on both counts, he accepts Amleth's word as if it were divinely inspired. Amleth proves to have such exceptional qualities that the king decides to follow through with the request of the message and put Amleth's guards to death. Amleth returns to his native land, liquidates his uncle's henchmen, and then Feng himself.

Shakespeare severs his ties with Saxo completely at the end of Book 3 because, in Book 4, Amleth "enters on a wholly new set of adventures which Shakespeare ... did not need" (Elton 400). The death of Koll at the hands of Horvendile provides a basis for Old Fortinbras' death at the hands of Old Hamlet. The deaths of Rosencrantz and Guildenstern duplicate the execution of Amleth's guards. Feng and Geruth are associated with the Danish royal couple Claudius and Gertrude. Amleth's killing of Feng is the precedent for Hamlet killing Claudius. Saxo's British king believes in divine direction, matching Horatio's proclamations that "Heaven will direct it."

Shakespeare does not incorporate the King of Britain's offer to Amleth of his daughter in marriage because it does not suit his purpose to complicate the relationship between Britain and Denmark. The Bard wishes to unite Poland and England in a common cause but Poland is mentioned only once in *Historia Danica*, in connection with the exploits of Starkad, son of Storwerk. Starkad has a powerful physique and slays the champion warrior of Poland, but Shakespeare could hardly use the defeat of a pre-eminent Polish warrior when he wishes to laud a pre-eminent Polish mathematician. Historians have debated whether Copernicus was Polish or German because his hometown lay in a region over which the King of Poland had some sort of suzerainty. Shakespeare might have created an indirect Polish connection through Germany, but a similar difficulty arises there because Germany's association to Denmark in Saxo occurs twice and, in both cases, the Danes emerge triumphant militarily. For this reason, Shakespeare invents Fortinbras and his foray into Poland.

Equivocation.

Historia Danica and *Hamlet* have in common the persecution of the young princes by their evil kings and the way in which each prince copes with his straits. Both princes use language whose meaning is open to interpretation. To the charge that he gives cunning answers, Amleth answers that he had spoken deliberately for he "wished to be held a stranger to falsehood." Accordingly, Amleth "mingled craft and candour in such wise that, though his words did not lack truth, yet there was nothing to betoken the truth and betray how far his keenness went." Most thought that his speech was "idle" yet it "departed not from the truth." Amleth's "jest did not take aught of the truth out of the story," for, although it seemed senseless, it "expressly avowed the truth" (Gollancz 107-27). The characterization of cleverness as madness acknowledges the merits of hypothetico-deductive reasoning, and, to this day, scientists are often depicted as "mad." Amleth practices the art of double-speak to survive, and Hamlet follows suit. Ambiguity promotes interpretive duality and, given the narrow-mindedness of the times, Shakespeare in all likelihood would use it to avoid censure.

Other Hamlets.

Saxo's history suited Shakespeare's goals by enabling him to call upon Amleth's effected disorder in order to address the transcendent topics of infinite space and the domain of the deities. In so doing, it appears that Shakespeare owes little, if anything, to other Hamlets. The lost play *Ur-Hamlet* was performed as early as 1589. Its author was possibly Thomas Kyd (1558-1594) or at least someone who modeled himself on Kyd, but its relationship to Shakespeare's *Hamlet* is unknown. *Histoires Tragiques* by François de Belleforest (1530-1583), translated into English as the *Historie of Hamblet*, follows Saxo closely. The subject of Hamlet's celebrated melancholia arises and there is some moralizing. Belleforest introduces the subject of divination via the belief structure of the British king who discusses Hamlet's supernormal insights in the context of magic and the Scriptures, but emphasizes a spiritual Heaven to the neglect of the physical heavens. *Der Bestrafte Brudermord* (Fratricide Punished) was first staged in Dresden in 1626.

It may have derived from both *Ur-Hamlet* and *Hamlet* or simply from an abridged version of Q1, and has a "hasty, stripped quality" that is "crude" and "farcical" (Hibbard *Hamlet* 373; Satin 383). It focuses on the obvious superficial themes of Hamlet's supposed madness and the revenge for the unnatural murder of King Hamlet at the hands of his brother. The murderer's name is Erico, and there is no astronomy of any significance. Hamlet fools the two nameless attendants that accompany Prince Hamlet to England and they end up killing each other.

According to Edwards (8-9), the textual problem of *Hamlet* is "of great complexity" owing to variations that "are not alternative versions of a single original text but representations of different stages in the play's development." It appears, however, that the mooted sub-text on physical cosmology and empirical methodology in Shakespeare's *Hamlet* owes nothing to other "Hamlets," with the possible and unverifiable exception of *Ur-Hamlet*.

Balderdash.

One explanation for the seeming insanity of the princes is that they fit the "balderdash syndrome" of the "almost correct answer," otherwise known as Ganser's Syndrome (Youngson 1-3). The ailment occurs among those who are not clinically mad but have something to gain from seeming so. It occurs predominantly among males who are threatened or abused, and among members of the armed forces. It is unknown whether Ganser's syndrome is a genuine psychiatric condition or simply a ploy in the struggle to survive. Sufferers "try it on" in order to be thought mad, the argument being (one supposes) that someone who prattles incoherently but is not overtly a threat to the state is unworthy of persecution or assassination. Hamlet is potentially a sufferer because Claudius menaces him, just as Amleth lived in fear of Feng. Hamlet states his strategy, "I perchance hereafter shall think meet / To put an antic disposition on," and he tells his mother that he is not insane but only "mad in craft."

In the modern world, professionals easily detect the Syndrome and, in the classic tales of Saxo and Shakespeare, the respective kings eventually penetrate the cover as well. So does Polonius: "Though this be madness, yet there is method in't," he mutters. Claudius calls on Rosencrantz and Guildenstern to discover why Hamlet "puts on this confusion ... and dangerous lunacy" and, within a few lines, Guildenstern follows Polonius in characterizing Hamlet's "true state" as "crafty madness." Hamlet's condition worries the royal court and, in 3.1, Polonius recommends exile or imprisonment, "To England send him; or confine him." Claudius knows that "Madness in great ones must not unwatched go."

The prince's melancholia and the king's paranoia make for a heady mix. The oppressors persist and, in 3.2, Rosencrantz warns Hamlet that he had better explain his condition or face the consequences, "You do surely bar the door to your own liberty if you deny your griefs to your friend." Claudius is worried as well, "The terms of our estate may not endure / Hazard so near to us." Matters come to a head as the usurper kings in both tales send their afflicted subjects packing and arrange for their executions in England. Hamlet's dread of "bad dreams" is fully justified for, until his task is complete, he must avoid the abysmal fate that awaits perpetrators of original thought. The Bard may have in mind the fate of the demi-god Balder who, in Saxo's Book 3, has bad dreams and is slain. If Balder's hopes are dashed, then so might Hamlet's.

Deceit.

Hamlet's modus operandi contrasts sharply with the deceitful methods of the king and his courtiers. Deceit first surfaces as Polonius and Reynaldo open Act 2. Polonius wishes to keep an eye on his son, Laertes, who has returned to Paris, so he gives his agent, Reynaldo, a few tips on how to gather intelligence. "By indirections find directions out," he tells him. He tells his daughter, Ophelia, that he has learned – somehow – that she has "been most free and bounteous" with her time spent with Hamlet and berates her zealously. "What is between you? Give me up the truth," he commands. When Ophelia admits that Hamlet had made tenders of affection for her, Polonius spits out "Affection? Puh!"

Most fathers would welcome a prince for a son-in-law, but not Polonius. From the outset, he works to scuttle the budding romance by twisting the fact of Hamlet's amorous interest into a symptom of lovesickness, his purpose being to label Hamlet insane. Polonius brings his explanation to the attention of the royal couple and is confident that he will receive a hearing. In 1.2, he had observed that Claudius knows that Hamlet is not quite himself and that Claudius had lectured Hamlet at length on the errors of his ways, so he initiates a campaign to discredit Hamlet by feeding the concerns that Claudius and Gertrude already have about him.

Polonius is fully committed to the king's cause, "I hold my duty, as I hold my soul, / Both to my God and to my gracious king." He tells the royal couple that he knows the cause of Hamlet's lunacy, but, when Polonius is briefly offstage, the queen wonders aloud whether Hamlet's alleged condition results from his father's death and her "o'erhasty marriage." The queen has offered a counter argument and the king decides to test it by "sifting," or questioning, Polonius in order to separate the wheat from the chaff of his opinion. Evidently, Claudius, Gertrude, and Polonius are in the dark and have no idea of the real cause of Hamlet's condition.

This exchange shows that Polonius sees no reason to test his theory, but the king is a bit more enlightened. Claudius is a cut above Polonius just as, by the second century AD, Claudius Ptolemy had progressed slightly beyond most of the philosophers of ancient Athens because he gathered data and incorporated them into his work. Ptolemy and his predecessor, Hipparchus, whom he much admired, followed in the footsteps of the Athenian, Meton (b.c.460 BC), who was the first "scientific" astronomer of Ancient Greece and was so-called because he made observations.

Polonius returns to the theme of Hamlet's insanity and presents a letter as evidence that he is mad for Ophelia. In defending his position, Polonius asks whether there has ever been a time that he had positively said that something was so when it proved otherwise. He says that he always gets to the truth even if he has to go to the center of his Universe. Polonius' geocentric worldview is primitive and he sees himself in a special relationship to its center where his core values reside (Edwards 2.2.157n; Goddard I, 407 orange he. Hamlet holds a man who is not passion's slave in his heart's core but Polonius seeks truth at the center of his own mind where, of course, he always finds it.

The idea that the controlling divinity of the World is centrally located dates at least to the Pythagoreans, but another view, extant in the sixteenth century, is that destiny is ordained from the opposite direction, from the heavens. Just as geocentricists believed in a simplistic interpretation of celestial appearances from a terrestrial point of view, so Polonius thinks that what he spies with his own eyes or thinks with his own mind is a faithful depiction of reality. Polonius seeks to impose his delusional thinking upon those around him and, by not seeking empirical verification of his political, social and scientific views, he weakens the very state that he seeks to protect. Polonius suffers from a dangerous methodological condition – the delusion of infallibility. Only reluctantly does he agree to a test of his theory, and then entangles his daughter in his mesh of iniquity. He suggests that they contrive an encounter with Ophelia as Hamlet walks in the lobby while he and the king lurk behind an arras to mark it. The king is barely a cut above his wayward counselor, for he agrees to the proposal. "We will try it," he says. The setup is worthy of Set, the God of Evil, who killed his brother Osiris and usurped his throne.

Possession.

The spymaster can get along without his son but is highly protective of his daughter. "I have a daughter – have while she is mine," he exclaims. The remark is puzzling, for Ophelia will always be his daughter. The repeated "have" suggests ownership. With a grim irony, Polonius could actually foretell Ophelia's death because, if she were to die, she would no longer reside in his household and he would no longer "have" her.

Polonius' possessiveness opens the door to further deceit. He forbids Ophelia to have further contact with Hamlet and, after Ophelia complains that Hamlet has frightened her, he asks her whether Hamlet is mad for her love. She replies that she does not know but fears that he is. Polonius promptly concludes, "This is the very ecstasy of love." Suddenly, Polonius loses his train of thought as if new thoughts were crowding out the matter at hand. "I am sorry," he says as he regains composure, and then he asks, "What, have you given him any hard words of late?" Ophelia replies that she did not and that, as commanded, she repelled his letters and denied him access to her. Polonius concludes unequivocally, "That hath made him mad." With that, the conjecture of 2.1.83 has evolved into the certainty of 2.1.108. It is as Pendleton (71) suggests in another context, "what enters ... as possibility ultimately coagulates as fact." Either Polonius falls victim to the fallacy of the Argument from First Cause ("after it therefore because of it") or he knowingly establishes a false causal connection between love and madness in order to fuel animosity toward Hamlet. Polonius has such an important position in government that he should not be prone to too many logical blunders, yet he is certain that denial of access to Ophelia has made Hamlet "mad."

What prompts Polonius to change the subject, and ask whether Ophelia has given Hamlet "any hard words"? According to the *OED*, the adjective "hard" can mean firm, unyielding, difficult to penetrate with the understanding, not easy to understand or explain, callous, hard-hearted, or unfeeling, cruel and harsh. All these meanings could apply to Ophelia's rebuff of Hamlet because breaking up is hard to do. Perhaps Polonius used the word "hard" because he is afraid that Ophelia will complain of his callousness. Could this be why Polonius is suddenly so distracted, and why, once reassured that all Ophelia did was rebuff Hamlet, he seems relieved?

One would think that a charge of madness would require corroboration, yet Polonius has no qualms about creating a circumstantial case. Ophelia's reply reassures Polonius and, immediately, he divulges what it is that he is sorry about. He regrets not having observed Hamlet more diligently but, evidently, has no regrets about his mistreatment of Ophelia. In deciding to "loose" her to Hamlet, Polonius speaks in terms more suitable to animals than a daughter. He uses her for selfish purposes, as if she were chattel. He could benefit from following Tycho Brahe's motto, *Non Haberi, Sed Esse* (not to have, but to be), because Ophelia has a right "to be" and her father does not "have" her. He meddles in the course of her maturation, creates conditions antipathetic to her life and leaps to conclusions that suit his own agenda. Polonius is obsessed with the desire to protect his daughter and will use any means to do so. Polonius is in the grips of a downward spiral of scurrility and irrationality. He cannot distinguish appearance from reality and, having dismissed the need to confirm theories, he suffers a geometric increase in error that is proportional to the state into which his mistakes have led him.

In 2.2, Hamlet warns Polonius to attend to his daughter, "Let her not walk i'th'sun. Conception is a blessing, but as your daughter may conceive – Friend, look to't." One commonplace of the time was that "the Earth is fertilized by the sun and conceives offspring every year" (Copernicus 26), and another was that maggots and the like are generated spontaneously from carrion bathed in sunlight. It is common knowledge that young princes have dalliance on their minds and, surely, Polonius does not need Hamlet to warn him of what comes naturally.

Divine Influence.

Both Saxo's *Historia Danica* and Belleforest's *Histoires Tragiques* portray the hero as one with supernormal insight, and the expectation is that Shakespeare would endow Hamlet similarly. This view finds ample textual support, for either Hamlet is remarkably prescient or his luck is simply out of this world. For all the evidence of Hamlet's ratiocination, divine influence plays an important role because, were it not for its apparently systematic effect on outcomes, some occurrences would otherwise strain credulity. Just as Shakespeare breaks down the old conceptual barriers to an infinite World, he goads his audience into thinking outside the box of the natural world.

Horatio's part is full of inconsistencies as well, as when, having "been absent at Wittenberg [Germany], he is able to inform the Danish soldiers about what is happening in their own country" (Edwards 1.2.176n). It is more likely that Shakespeare is writing consistently by making Horatio a conduit in a supernatural drama. Horatio says, "My lord, I came to see your father's funeral." The funeral of old Hamlet has long since taken place, so perhaps Horatio announces that he came to see the funeral of Claudius. At this early stage, however, Horatio has no reason to suspect that the death of Claudius is imminent.

When Hamlet ponders whether the Ghost will reappear, Horatio says, "I warrant it will." Starting in the fourteenth century, the verb "warrant" has meant to "guarantee as true" (*OED*). Shakespeare used the word with that meaning thrice in *The Merry Wives of Windsor* of 1598 and, in *Henry VI part 1* of 1591, in the sense of "to promise or predict as certain." Horatio assures Hamlet that the phantom will walk again, but how can he be so sure?

Horatio fears that the Ghost might deprive Hamlet of his powers of reason by drawing him into madness. Horatio anticipates the strategy of madness that Hamlet proclaims only in the next scene. Horatio pronounces, "Heaven will direct it," suggesting that events are supernaturally controlled. In the next scene 1.5, Hamlet tells him, "There are more things in heaven and earth, Horatio, than are dreamt of in your philosophy." Horatio does not foresee everything; for example, he is unaware of the seemingly miraculous intervention of the North Sea pirates. Marcellus responds to Horatio's pronouncement that heaven will direct events by saying, "Nay, let's follow him," where "nay" means "let us not leave it to Heaven, but do something ourselves" (Jenkins 1.4.91n).

After the Ghost reveals details of Old Hamlet's murder, Hamlet mutters, "O my prophetic soul." Edwards (1.5.40n) wonders whether Hamlet has "guessed" a truth, but Hamlet wonders whether he is possessed of supernaturally endowed foresight. Jenkins (1.5.41n) suggests that Hamlet's "prophetic soul" refers to "divination not of the murder ... but of his uncle's true nature." This makes sense in a superficial literal reading but can refer also to Claudius' role as chief advocate of bounded geocentricism.

Hamlet says that he does not know why he has lost all his mirth, yet he goes on to complain about deficiencies at Elsinore that, in theory, could be the cause and that should be obvious to him. In 2.2, before he learns that the king has designs on his life, he tells Polonius, "You cannot sir take from me anything that I will more willingly part withal; except my life, except my life, except my life." Hamlet knows that he must survive long enough to complete his work. Hamlet speaks from a metaphysical script because he knows "that the whole supernatural world of good and evil lies behind his revenge," which is "instigated by heaven in its war against the workings of hell, visible in Claudius's achievements" (Edwards 2.2.537n).

Hamlet asks the touring thespians whether they can play *The Murder of Gonzago* and it so happens that they can. Either they have an enormous repertoire or Hamlet is lucky again. Before the Players' play commences, Hamlet shifts the conversation to the time when Polonius played Julius Caesar "i'th'university" and was killed by Brutus. Polonius is quite matter-of-fact about this, but the irony is that Polonius unknowingly foretells his own death as well as the death of the despot, Claudius.

Hamlet regrets killing Polonius and excuses himself because he is not completely in command of his own actions. He knows these have support from on high:

> I do repent; but heaven hath pleased it so,
> To punish me with this, and this with me,
> That I must be their scourge and minister.

Hamlet repents in the physical world but, allegorically, he is to perform the honorable duty of ridding the world of pedantry.

Hamlet knows that Rosencrantz and Guildenstern will accompany him and that they bear sealed letters for "the mandate." Hamlet cannot know all this beforehand. A popular explanation is that Hamlet accidentally overheard the king's planning but, barring the time when Hamlet overhears the king praying, the script says nothing about him eavesdropping on the king. If Shakespeare is not a slipshod dramatist, then it is likely that Hamlet is the beneficiary of metaphysical communication.

Hamlet asks his mother, "I must to England, you know that?" and she replies, "I had forgot." Gertrude's reply implies that she benefited from subliminal communication as well, but has a bad short-term memory. Perhaps, she is excusing herself on the grounds that, if he knows of his scheduled departure, she ought to know of it as well. Another possibility is that "to forget" means "to give no thought to" (Crystal and Crystal 184), suggesting that she had not even considered the possibility of Hamlet's exile.

Hamlet resolves to trust Rosencrantz and Guildenstern as he would adders fanged. He will delve one yard below their mines and blow them to the Moon. The *OED* uses this passage to illustrate the meaning of "delve," which means to "labour with a spade," or, simply, to "dig". One can well imagine why Hamlet, the pioneering son of the old digger mole, blows the guards to the Moon because he knows that the model they personify is as blemished as that destination. By native talent and the grace of higher powers, Hamlet will execute a twin killing, ridding the field of the two-component hybrid model and its two-fold personification (see Chapter 8).

The king says that Hamlet must sail for England "with fiery quickness," ostensibly for his own good. "For England?" asks Hamlet. "Ay," replies Claudius. "Good," says Hamlet, implying that the king's order is to his liking. "So it is if thou knew'st our purposes," mutters Claudius who thinks that Hamlet fails to appreciate his true intent, but Hamlet knows more than he lets on. "I see a cherub that sees them," he says, meaning that he knows that "heaven is watching" (Edwards 4.3.45n).

In 5.2, Horatio wonders how Hamlet re-sealed the forged commission. Hamlet explains that "even in that was heaven ordinant" for he had his father's signet in his purse. Pirates extricate the prince from premature entanglement with Britain and return him to Denmark. Old Hamlet was once a pirate and it looks as if his ghost is paving the way behind the scenes. The *deus ex machina* enables Young Hamlet to get on with his supernaturally directed task.

Hamlet explains that the pirates were "like thieves of mercy" who "knew what they did." How did the pirates know? Sources indicate that the "paradox of thieves showing mercy is wittily expressed by applying to thieves a phrase more commonly used of angels," whose mercy, moreover, "was calculated" and, in return, Hamlet is "to do a good turn for them" (Edwards 4.6.17-18n; Jenkins 4.6.19n). Jenkins (4.6.20n) believes that the "good" in the phrase "a good turn" is "superfluous" and "enfeebling," but the quid pro quo is nothing less than reframing the Universe. In the absence of this explanation, the rescue easily becomes the butt of satire because the real reasons for the pirates' actions seem so obscure.

Just before Hamlet tells Horatio of these miraculous occurrences, he refers to divine will, "There's a divinity that shapes our ends, / Rough-hew them how we will." Horatio replies, "That is most certain." According to Edwards (5.2.10-11n), "there is a higher power in control of us, directing us toward our destination, however much we have blundered in the past and impeded our own progress. Hamlet feels the guiding hand of heaven in his own impulsive and unpremeditated actions, after the failure of his own willed efforts." In Greek mythology, Apollo, as Theos, resembles as much a law of nature as he does a god, for that remote deity "accompanies the action on the divine plane [and] signifies that what happens below is the working of universal law" (Kitto 74-5).

In 5.2, Claudius rigs the odds to give the impression that he supports Hamlet in a swordfight with Laertes. In accepting, Hamlet willingly risks life and limb on behalf of one whom he has vowed to assassinate. This does not make much sense. Horatio believes that Hamlet is incapable of besting Laertes and offers to excuse him, but Hamlet refuses, "Not a whit, we defy augury. There is special providence in the fall of a sparrow. If it be now, 'tis not to come; if it be not to come, it will be now, yet it will come – the readiness is all." Hamlet speaks fatalistically of his own death for he senses that a superior power is in charge and he will not defy the script by letting Horatio make excuses for him. He likens himself to a sparrow whose life is in the hands of a "special providence" and resigns himself to whatever fate has in store for him. Edwards (5.2.192-3n) explains that, "All occurrences show God's immediate concern and control, and he [Hamlet] will therefore accept the circumstances which present themselves and not try to avoid them." Disbelief is easier to suspend if we presume that an all-knowing and benevolent deity is operating behind the scenes.

The evidence warrants a non-literal interpretation whose sub-text concerns the supernatural world. Shakespeare leaves no doubt about where his sympathies lie as he creates a classic confrontation between new and old, right and wrong, and good and evil. He contrasts the methods of deceit practiced by the Elsinore establishment, with the enlightened methods employed by Hamlet and, of all the ways that he uses to advance the plot, *The Murder of Gonzago* is the pivot about which *Hamlet* turns.

CHAPTER 7: THE FULCRUM

The best-laid plans o' mice an' men Gang aft a-gley.
Robert Burns

For a limited time the Ghost has permission to visit its old haunts and, in 1.5, uses the time to instruct Hamlet to avenge his father's foul and most unnatural murder. At first, Hamlet assumes that the apparition is honest, but then he worries that it is a devil and a goblin damned bringing blasts from hell and bent on snaring Hamlet's soul. If the Ghost is lying about Old Hamlet, it may also mislead Young Hamlet when it instructs him not to contrive against his mother. Hamlet remembers that the Ghost referred to Gertrude as a *seemingly* virtuous queen, but even if she bedded one brother while betrothed to the other, surely she would not stoop to murder to escape her marital vows? The prince is a practitioner of sound methodology and does not leap to conclusions, but avails himself of the divine gift of reason to decide the matter.

The Attendants.

The royal couple wish to help Hamlet recover and invite two of his friends, Rosencrantz and Guildenstern, to visit and discover what ails him. The courtiers locate Hamlet and engage him in conversation. Hamlet bemoans his lot. Nothing seems right to him, not even man who is that magnificent creation, the paragon of animals and quintessence of dust. Hamlet expresses this sentiment by saying, "Man delights me not." His remark elicits smirks from the two trusties, prompting Hamlet to add, "no, nor woman either, though by your smiling you seem to say so." At the very outset, the king's agents engage in presumptive theorizing of a sort characteristic of pre-Socratic thinking. They leap to a conclusion unwarranted by evidence, which puts them and Polonius squarely in the camp of the methodologically challenged.

Rosencrantz denies that his reaction is inappropriate. "There was no such stuff in my thoughts," he says, tripping over his tongue. He fails to define what he means by "stuff," and Hamlet cannot let the insinuation drop. He asks, "Why did ye laugh then, when I said man

107

delights not me?" Rosencrantz deflects the question by answering, "To think, my lord, if you delight not in man, what lenten entertainment the players shall receive from you." Rosencrantz tries to shift the topic of conversation away from his gaffe by referring to the theatrical custom of stage actors being exclusively male. He pretends that he thought that Hamlet meant that theatrical performances by men no longer pleased him. He presses his advantage by adding that Hamlet would give such an actor a cool reception (a "lenten entertainment"). Rosencrantz has wrenched his foot out of his mouth at the cost of changing the subject to "the players." "We coted them on the way, and hither are they coming to offer you service," he explains.

Without prior mention of actors, Rosencrantz has said that "the players" are on their way, and, equally, Polonius, not having been privy to the previous discussion, announced that "the actors" have arrived. In both cases, the definite article must refer to a particular group of players that they knew about beforehand. The inference is that the summons for the thespians came from the royal court and, since both Polonius and the courtiers are in the know, the likely source of the summons is the king.

Without missing a beat, Hamlet replies, "He who plays the king shall be welcome," implying that he "is prepared to honour one pseudo-king with as much seriousness as another" (Edwards 2.2.298-9n). Thereby, Hamlet identifies the king and he promptly speaks ill of him, "The clown shall make those laugh whose lungs are tickle o'th'sere." Hamlet speaks of those whose sense of humor is easily tickled, where a "sere" is a catch affecting the trigger-mechanism of a gun.

The banter careens on at breakneck speed. Either Hamlet can think faster than greased lightning or he responds as if his speech is scripted. He will let the courtiers wallow in their muddled thinking and on the sly, will use the actors to further his own agenda. We do not yet know his plans, but Hamlet predicts that a certain regal clown will react to a dramatic "sere" as if he were a piece that goes off half-cocked. The usurper thinks he is a big gun and Hamlet plans to fire him.

The Test.

If a beneficent spirit hovers sight unseen, so does Satan. Prior to uttering the enigmatic lines on madness and wind directions (see

Chapter 11), Hamlet seizes the opportunity to announce his strategy. He will use a stage play to test the Ghost's honesty and, to that end, he makes an epistemological vow, "The play's the thing / Wherein I'll catch the conscience of the king." A play performed on stage is to Hamlet as a laboratory experiment is to a scientist. Hamlet will test the Ghost's directive with a play-within-the-play to see whether it triggers a response from the king.

Hamlet's methods are the antithesis of his befuddled cohorts'. He follows steps (1)-(4) listed in Chapter 6, beginning with knowledge accumulated from observations and experiences gleaned in the real world of Elsinore and Wittenberg, followed by the hypothesis that the Ghost is honest and, therefore, that its report on the guilt of Claudius is true. The test relies on the received fact:

> That guilty creatures sitting at a play
> Have by the very cunning of the scene
> Been struck so to the soul, that presently
> They have proclaimed their malefactions.

Hamlet's methodology is rare, for he will conduct a test of the truthfulness of a supernatural being with the help of events in the natural world.

The players' play occurs halfway through *Hamlet* and is the fulcrum about which the actual play turns (Wilson *Happens* 137 *ff.*). Hamlet and Claudius square off, the former seeking to advance the New Philosophy as the latter defends the Old. The foes oppose one another both in natural and supernatural space, with Hamlet believing that his cause has the blessing of spirits on high and Claudius believing in the doctrinal certainty of Christianized Aristotelianism. A fight to the finish will ensue both on stage and in the imperceptible dimensions of supernatural space. In effect, the play-within-the-play becomes an allegory-within-the-allegory and requires examination and interpretation in that light.

The Script.

The royal couple seeks to heal Hamlet from illness contracted at Wittenberg and restore him to geocentric health. They want Hamlet to follow their advice. He should grieve over the death of his father for a month or so, curtail his Wittenberg education, climb aboard the geocentric bandwagon and enjoy life in the warm embrace of Elsinore. Since they hope to win him over, they offer all the support they can and plan to attend the performance. In order to please the royal couple, the actors must not get into an argument with Hamlet because their job is to restore his health. Claudius has warned the courtiers that something saddens the prince beyond the death of his father and we expect that the entertainment that is intended to improve Hamlet's humor is actually a prescription for a dose of geocentric medicine.

Hamlet asks whether they can play *The Murder of Gonzago*. Luckily, they can, and Hamlet says that he will suggest a change in the script, "You could for a need study a speech of some dozen or sixteen lines, which I would set down and insert in't." Hamlet associates his emendations only with "speech." When alone, he says that his intent is for the actors to, "Play something like the murder of my father / Before mine uncle." His alteration will only resemble, be "something like," the circumstances of his father's murder. The inexactness recurs when Hamlet tells Horatio that one scene of it "comes near the circumstance" of Old Hamlet's death. Hamlet's strategy is puzzling, for why change the text so that it only resembles the crime when a precise re-enactment would be more likely to elicit a reaction of guilt? Hamlet instructs Player I how to "speak the speech ... trippingly on the tongue" and allow words and action to suit one another, but he says nothing about actions. Hamlet plans to test the conscience of the royal couple by using "one speech," but says nothing about stage directions.

Ambition.

Hamlet theorizes that if Claudius takes the altered play in stride, he is probably innocent and the specter is damnable. Hamlet tells Horatio, "Observe my uncle." When the royal party arrives, Claudius inquires after Hamlet's health, "How fares our cousin Hamlet?" Hamlet pretends to misunderstand the meaning of "fare" and replies that his "fare" is "the chameleon's dish" which is "air, promise-crammed." Chameleons were once believed to thrive on air and Hamlet likens himself to one because Elsinore offers nothing substantial to feed his soul. His only nourishment is wind pudding and air sauce, commonly called "hot air." If Hamlet must eat air like a chameleon, then he may have another characteristic of that Old World group of reptiles – he could change the color of his stripes. After the players' play ends and the courtiers rebuke Hamlet for causing a scene, the prince explains why he might don a cloak of another color – he lacks "advancement." Rosencrantz, who had already suggested that Hamlet is ambitious, reminds him that he is in line to succeed Claudius. The dietary theme persists as Hamlet refers to the proverbial horse that starves while waiting for the grass to grow.

The Players' Play.

The play-within-the-play commences with a dumb show that faithfully mimics the Ghost's account of the alleged crime. Hamlet had spoken unflatteringly of "inexplicable dumb shows and noise" but the players stage a dumb show anyway because their mimed re-enactment is part of the geocentric prescription. The players did not argue with Hamlet because, to them, Hamlet's opinion about dumb shows is as worthless as his opinion about the Cosmos.

The dumb show begins and Player King and Player Queen show their love for one another. He lies down and sleeps, and she leaves him. An unidentified man enters, removes the crown from the Player King's head, and pours poison in his ear. The Player King dies, and the murderer woos the Player Queen who soon accepts his love. Surely, Claudius could not fail to see that the unnamed murderer represents him, but, oddly enough, he does not bat an eye. Some think that he is preoccupied and does not see the dumb show. Others believe in the so-called "second tooth" theory that, initially, Claudius has the fortitude to withstand the implications, only to cave in later. The *Hamlet* script supports neither explanation. Instead, suppose that, beneath Claudius' human facade, lies a metaphysical persona, devoid of human feeling and incapable of emotion. Just as the allegorical sub-text transcends the literal story line, his otherworldly self dominates and Claudius cannot react guiltily, even to a precise rendition of his crime.

Dichotomy.

Shakespeare is at pains to reveal that Claudius does indeed possess a conscience and, thus, a human persona. In the scene immediately preceding the players' play, Polonius comments, ironically, that, "'Tis too much proved, that with devotion's visage, / And pious action, we do sugar o'er / The devil himself." Aside, Claudius agrees. "Oh, 'tis too true," he says, "How smart a lash that speech doth give my conscience!" Then in the scene immediately after *The Mousetrap*, Claudius admits, prayerfully, that his lese majesty is rank and smells to heaven.

At all other times, Claudius plays the role of a zealot in defense of the geocentric faith and is impervious to arguments from the real world. He must defend cosmic virtue against the modern blasphemers because, as Guildenstern remarks, the preservation of geocentricism is a "holy and religious" cause. Claudius is on a divine mission. His unfeeling persona knows that the end is infinitely worthy and that any means justifies the end. From the start, Claudius needed a seat of power and achieved it by killing Old Hamlet and usurping the throne. To a self-righteous geocentricist in the metaphysical world, regicide is all in the day's work, and remorse is not a factor. The dumb show merely recounts what Claudius sees as an honorable means to a virtuous end and his unfeeling self is impervious to the implications of the mimed re-enactment. Even if his metaphysical self could emote, he would act in a conceited manner and be not in the least guilt-ridden.

For their part, the thespians have no qualms about staging the pantomime because, as Claudius' hirelings, they know that the king's supercilious self insulates him from the real world. Furthermore, the thespians feel free to butter up Claudius by catering to his cause and to accomplish, literally and figuratively, what he wants them to do, which is to administer a geocentric nostrum to the wayward prince. The Ghost has opened Hamlet's eyes to two modes of existence, so Hamlet knows the futility of confronting the king's metaphysical self with physical evidence. No matter whether Claudius is guilty or innocent, the confrontation will evoke the same reaction – none. In order to trap the king, Hamlet must test his other self, his human ego, and Hamlet knows that he must alter the *Gonzago* play accordingly.

112

Mischief.

Ophelia asks Hamlet the meaning of the dumb show, sensing that it "imports the argument of the play." She sees the miming as portentous and is puzzled because dumb shows generally do not divulge key aspects of the action to follow. Hamlet replies, "Marry this is miching mallecho, it means mischief ... the players cannot keep counsel, they'll tell all." Previously Hamlet had entrusted them to adopt the changes he had requested, but their staging a dumb show expressly against his wishes has tipped him off. The thespians are potential troublemakers because they are in the thrall of the king and will not follow Hamlet's bidding. They must advance the king's agenda both literally and figuratively, so will warn Claudius' emotional being of Hamlet's designs and, simultaneously, laud what we in the literal world call the crime of Claudius.

Hamlet predicts that the thespians will "tell all," but what "all" do they intend to tell? Context supplies an answer. Hamlet mutters "miching mallecho" immediately after the dumb show ends. If the players were so emboldened as to depart from the original script by inserting a dumb show, then they would surely feel free to compromise the spoken lines that Hamlet has asked them to insert. In matters of such import, Hamlet cannot take any chances. Lest they fail to execute his instructions, he must devise a new strategy to replace the one that the players already know about. He must build a better mousetrap. Hamlet's plight is immediate because the oral show is about to begin and he has no opportunity to give the players a dressing down or instruct them anew.

Oral Show.

Hamlet is equal to the task regardless of thespian mischief. The oral show commences with the entrance of the Prologue whose three lines of jingling doggerel give no clue to the drama to follow. The Player King and Queen then converse and allude to the duration of their marriage:

> Full thirty times hath Phoebus' cart gone round
> Neptune's salt wash and Tellus' orbèd ground,
> And thirty dozen moons with borrowed sheen

About the world have times twelve thirties been,
Since love our hearts, and Hymen did out hands,
Unite commutual in most sacred bands.

Sohmer (*Mystery* 235) notes that the player couple's wedding day was 360 lunar months plus 30 solar days ago, or 29 years and 69 days ago, where a synodic month is the period of a lunation equal to 29.5 solar days. Thus the opening lines hint that the Player King and Player Queen are Hamlet's parents and imply that Hamlet's parents had been married almost thirty years at the time of Old Hamlet's murder, yet later it turns out that Hamlet has turned thirty, implying that he is illegitimate. Claudius and Gertrude must know Hamlet's age and, for different reasons, could not fail to wonder at the implications of events on stage.

Player King predicts that he must leave Player Queen soon and she vows fidelity to him even after death. She implicates wives in the crime of murder and not second husbands. The Player King's response is prolix, as if to give everyone a chance to gather their wits and ponder the implications. He ends by expressing skepticism of Player Queen's fidelity, "So think thou wilt no second husband wed, / But die thy thoughts when thy first lord is dead." However, she vows, "If once a widow, ever I be wife." The newly-married Gertrude could not fail to feel a prick of conscience.

The Player Queen's talk of wives killing husbands hints that maybe it was Gertrude who killed Old Hamlet, and not Claudius. Hamlet must discover his mother's innocence or guilt, so he runs another test. He asks Gertrude how she likes the play, and she replies that she thinks the Player Queen "doth protest too much." Hamlet mocks her, "Oh but *she'll* keep her word" (emphasis added), meaning that the Player Queen intends to remain faithful even if the real queen, Gertrude, did not. The culture of the time frowned on widows remarrying, even though many did.

Mention of murderous wives catches the attention of Claudius, prompting him to ask Hamlet "Is there no offence in't?" Claudius depends on Gertrude, perhaps even loves her, and here he shows his human side by exhibiting concern for her. For her part, Gertrude is a wife and mother and has no role in the metaphysical sub-text. Her character is relatively straightforward and, like her name, is essentially unchanged

114

from the Saxo tale (see Tables 2, 3). She belongs to the real, literal world, which is why the Ghost instructs Hamlet to let Heaven judge her as it would anyone who seeks passage through the pearly gates. Her normalcy explains why Hamlet sees the Ghost in her chambers and she does not. Some people live their entire lives without ever seeing a ghost, so it is reasonable to suppose that a spectral property permits selective visibility. The Ghost would accomplish nothing by revealing itself to her, other than to scare her when she sees her dearly departed floating around in her bedroom.

Table 2

IDENTIFICATIONS IN SHAKESPEARE'S HAMLET

SHAKESPEARE	SAXO	IDENTIFICATION
Claudius	Feng	Ptolemy
Gertrude	Geruth	
Rosencrantz Guildenstern		Tycho Brahe
Polonius		Robert Pullen
Reynaldo		John Reynolds
Ophelia		
Laertes		Thomas Harriot
Osric		Walter Raleigh
Ghost	Horvendile	Leonard Digges
Hamlet	Amleth	Thomas Digges
Barnardo		Bernardus Silvestris
Francisco		Francesco Petrarca
Marcellus		Marcellus Palingenius

Table 3

IMPUTED ROLES IN THE PLAYER'S PLAY

FORMAL ROLE	PLAYERS' *GONZAGO*	HAMLET'S *MOUSETRAP*
Player King	Old Hamlet	Claudius
Player Queen	Gertrude	Gertrude
Murderer	Claudius	Hamlet

Claudius cannot function without Gertrude. "I could not but by her," he says. He is vulnerable because he cherishes her in physical space where the normal conditions of human existence prevail and Hamlet uses this vulnerability to penetrate his supersensible facade and reach his emotional being. The potential threat to his new bride distracts him and he does not see the trap that lies ahead. Hamlet lapses into bafflegab and assures the king that the play imports no offence, "No, no, they do but jest, poison in jest, no offense in the world." Claudius asks for the name of the play, and Hamlet answers, "The Mousetrap." Hamlet will set a trap and Claudius will "in jest" the poison.

Lucianus.

In the dumb show, an attentive Claudius would have seen that the Player King and Player Queen represented Old Hamlet and Gertrude and that the unnamed murderer was Claudius. For reasons stated, Claudius is undismayed and Hamlet suspects that he is about to get a double dose of medicine as the players enact an oral equivalent. Claudius will not flinch at the *telling* any more than he did to the *miming*, but Hamlet knows what to do. Player King and Player Queen are on stage and a player enters who Claudius anticipates represents himself. Suddenly, Hamlet announces, "This is one Lucianus, nephew to the king" (see Table 3).

Hamlet and Ophelia resume their chatter but the king does not silence them lest this impairs his recovery. Restoring his respect for geocentricism is, after all, the purpose of the whole production and, in any case, no one takes the prattling prince seriously because everyone knows he's nuts. Nevertheless, there is method in Hamlet's madness. When Hamlet says that Lucianus is "nephew to the king," he means that Lucianus represents the nephew of the Player King, but, in the real world, the only king with a nephew is the *new king*, Claudius. This gets the attention of Claudius who cannot help but wonder whether Lucianus represents Hamlet rather than himself. Hamlet's suggestion creates new identities for Lucianus and the Player King, and only the identification of the Player Queen with Gertrude remains the same.

116

Revenge.

After chitchatting with Ophelia, Hamlet blurts out, "So you mistake your husbands." Leaving aside any implication that mistaken husbands might have for Ophelia's chastity, we see that Hamlet is drawing attention to his mother's incest because Gertrude is "confusing" her husbands. Hamlet continues, "Begin murderer. Pox, leave thy damnable faces and begin. Come, the croaking raven doth bellow for revenge." The "croaking raven" refers to an old chronicle play, *The True Tragedy of Richard III* of 1594, which is all about revenge. Ominous signs of ghosts that come gaping for revenge against a murderous usurper are the Sun shining hotly, the eclipsing Moon, retrogressing planets and stars that turn to comets. Comets refer to supposed omens, like the comet of 1577, and possibly, to novae, which European chroniclers called comets, making the New Star of 1572 another pretext for revenge.

Hamlet calls Lucianus a murderer, but of whom? The opening words of Lucianus answer the question because they paraphrase the story that the Ghost had previously relayed to Hamlet. In addition, Lucianus pours poison in the ear of the sleeping king who, supposedly, is now the murderer's uncle. In case Claudius does not take the bait, Hamlet makes the point abundantly clear, "A poisons him i'th'garden for's estate. His name's Gonzago ... You shall see anon how the murderer gets the love of Gonzago's wife." With that, Hamlet springs the trap. The king loses composure, Ophelia announces, "The king rises." Claudius orders the lights to be switched on and, thereby, ironically, anticipates the dawning of a new age of enlightenment.

Before the entire court, Lucianus (Hamlet) has poisoned Gonzago (Claudius) in the same way as Claudius killed Old Hamlet, so that Claudius is "simultaneously confronted with the image of his crime and the threat of its avenging" (Jenkins 3.2.248n). Hamlet's *coup d'theatre* has caught Claudius off-guard and has triggered a response that has the appearance of guilt. The net result is that Claudius knows that Hamlet knows the true nature of his involvement and, for the first time, perceives Hamlet as a direct threat to his life. His course of action is clear – kill, or be killed.

Claudius strives not to die because, as the two courtiers point out at the start of the next scene 3.3, if he were to, the country would lose its center, the "lives of many" would be imperiled, the wheels of government would grind to a halt and, in the midst of the chaos, Hamlet would ascend the throne. The holy cause of bounded geocentricism would perish, the World would lose its nominal center and the wheels of the Cosmos would fall off.

Hamlet and Horatio exult at the success of the experiment and agree that the talk of the poisoning in the garden did the trick and not the mimed re-enactment. The test is over, Claudius is guilty and the Ghost is honest just like the rest of Shakespeare's visitors from the spiritual world. Hamlet's test is a great deal more sophisticated than any devised by the geocentricists, as expected of a dangerous lunatic who espouses both the empirical method of scientific inquiry and the utility of stage-plays.

The Altered Lines.

In Q2, the two opening statements by the Player King and Player Queen comprise 16 lines, which could be the "dozen or sixteen lines" that Hamlet inserted. These lines set the stage for future action but, in themselves, do not test the king's guilt. Hamlet probably did not supply these lines and, in any case, F1 gives the Player Queen two more, 3.2.152-3, bringing the total to 18, which exceeds the formal upper limit.

Hamlet's proposed insertion is an insoluble problem (Edwards 2.2.494n) because we never learn what textual changes he had originally ordered. The fact that Hamlet's oral interference did the trick suggests that his spoken words served in lieu of at least some of his proposed script. One possibility is that Hamlet assigns Lucianus the six opening lines that paraphrase the Ghost's tale, 3.2.231-6, and then proceeds to pre-empt the players' pro-Claudius plans. He does this in an unusual and unexpected way – by becoming a player himself. The additional lines that redefine the meaning of the players' play are 3.2.210 (Madam, how like you this play?), 3.2.212 (Oh but she'll keep her word), 3.2.221 (This is one Lucianus, nephew to the king), 3.2.228-30 (So you mistake your husbands ... bellow for revenge), and 3.2.237-9 (A poisons him i'th'garden ... Gonzago's wife). With the opening six lines of Lucianus, these amount to 15, or 13 if two lines referring to the croaking raven (3.2.229-30) are omitted as unessential. Perhaps it is coincidental that these 13 to 15 lines happen to fall in the range of Hamlet's original estimate of 12 to 16.

Harmony.

After the players' play ends, Hamlet calls for music, which signals a return to Pythagorean harmony. The courtiers arrive and Rosencrantz threatens Hamlet with incarceration. Players enter with recorders, and Hamlet asks Guildenstern, "Why do you go about to recover the wind of me, as if you would drive me into a toil?" Hamlet likens himself to a quarry, which, on scenting the hunter, flees with the wind only to be trapped in the net that the hunter has cunningly placed beforehand. Here, though, the net has a hole in it and the stratagem fails (Jenkins 3.2.336n, 337-8n, 341n).

Hamlet invites Guildenstern to play upon one of the pipes. "I have not the skill," says Guildenstern. Hamlet tries to teach him how to cover the holes of the chanter. "Govern these ventages with your fingers and thumb, give it breath with your mouth," he says. Hamlet's rebuke – "'Sblood, do you think I am easier to be played on than a pipe?" – is noteworthy. The oath "'Sblood" refers normally to the blood of the Savior, but could also refer to real bleeding from Tycho's facial wound. The imagery persists through identification of "mouth" as a source of air for people with stuffy noses. The origin of the discrete musical notes anticipates Kepler's Harmonic Law, which relates the separate planets' chief orbital dimension and their frequency of oscillation much as the pitch of a pipe varies with its dimensions.

CHAPTER 8: ALLEGORY

Out of sight out of minde, this may run right
For all be not in minde, that be in sight.
John Heywood

Allegory caters to the ideals of attainable knowledge and political unity in a moral and fundamentally theological context. It uses elaborate symbolism and has levels of meaning deeper than literal, making it a suitable means of relating Appearance and Reality. The technique of saying something in such a way as to convey non-literal meaning is especially useful when it comes to dealing with sensitive issues like the overthrow of corrupt regimes and false cosmologies.

Identity.

Hamlet opens on the Guard Platform of Elsinore Castle, named for Kronborg Castle in Helsingør on the northeast coast of Denmark. The Castle Platform is "a high terrace for mounting guns and keeping watch" (Edwards 1.1.0 SDn). It lies at the boundary separating outer and inner space, at whose extremities, according to the two-sphere model of the Universe, lie Heaven and Hell (Edwards 6). Garber (470-1) notes that boundary conditions are interesting in many contexts, and posits three in *Hamlet* that are "parallel to and superimposed upon one another." To these we add a fourth – the Universe's.

The play begins with, "Who's there?" Fourteen lines later Francisco repeats the question, signifying that identity is an issue. Suspicion falls upon the supernatural visitor, which, despite its otherworldliness, is nevertheless a bona fide character about which two-thirds of the lines of the first act are concerned. Barnardo and Marcellus have seen this visitor for two nights running, and Marcellus has gone to find Horatio so that he can see it too. Marcellus and Horatio identify themselves upon entering and Marcellus says that Horatio dismisses the accounts of the apparition as fantasy. Barnardo recalls that the Ghost appeared at one o'clock in the morning when the star "that's westward from the pole" was in the same position as it is now. The Ghost appears as if on cue and Marcellus says, "Thou art a scholar, speak to it Horatio." Horatio addresses the illusion imperiously, "By heaven I charge thee speak," and again, "Speak, speak, I charge thee speak!" To no avail. Barnardo comments that the apparition resembles the former king, Old Hamlet, at a time of war when he combated "ambitious Norway" and "smote the sledded Polacks on the ice." The specter reappears, and Horatio again implores it to speak, but it remains mute. He tries commanding it, "Stay and speak!" again without success. "'Tis gone," says Marcellus.

The epistemological component of the New Philosophy is operational from the start because Barnardo and Marcellus make observations, formulate a hypothesis and, by watchful waiting, conduct a test. The Ghost appears but the experiment is only partially successful because the further expectation that the Ghost will communicate with Horatio, fails. The spirit makes its intentions clear but again declines to speak, encouraging more observations and/or a better theory. Abetted by Horatio, Marcellus tries intimidation, but that does not work, either. After a short debate, Horatio, Marcellus, and Barnardo realize what they should have seen earlier, that they should seek out Hamlet, for surely a specter that resembles his father will speak to him. The modified theory proves successful. These elementary methodological lessons are such an advance over the method of untested assumption that its lesson must be supernaturally induced.

Tycho's Supernova.

Roth argues that the star "that's westward from the pole" is Alderamin, the brightest star in the constellation Cepheus, which, owing to Precession of the Equinoxes, will become a new North Pole Star in five thousand years. Sohmer (*Mystery* 219-23) argues for Deneb, the brightest star in Cygnus, whose name derives from the Arabic meaning "tail" but which Christians see as the head of the Northern Cross, standing erect above the western horizon at Christmas-time. These suggestions co-exist in their own interpretive frameworks with the suggestion that the star is the New Star of 1572 in Cassiopeia (Olson, Olson, and Doescher).

Shakespeare lets the apparition herald what Horatio describes in terms descriptive of what we know today as the explosive quality of Supernova stars. He opines that the celestial apparition bodes some strange eruption to "our state." Horatio does not say whether he has a political or cosmic state in mind. He could mean that he does not know in which particular area to concentrate his thoughts, so, on "taking a wide view" and allowing all possibilities (Edwards 1.1.67-8n), Horatio allows the New Star to serve as an omen of change in both terrestrial and celestial affairs. *Hamlet* leaves unresolved the physical nature of the eruptive star, as do Tycho Brahe and Thomas Digges in 1573. Fast-forwarding to 1921, we see that the so-called Great Debate on spiral nebulae and the Milky Way (Shapley and Curtis) also left unresolved the question of the physical nature of Novae. Even at that relatively late date, nobody knew that Novae and Supernovae are different phenomena, the former flaring to a peak power over 10,000 times that of the Sun, and the latter being 10,000 times brighter even than that (Whitney 216).

The fact that astronomers call the New Star of 1572 "Tycho's Supernova" biases modern readers into believing that Tycho's was the definitive work on it. Thomas Digges worked extensively on it, yet at the turn of the twentieth century, his *Alae seu* was regarded as "an astronomical treatise of no great importance" (Berry sects. 95, 100). Johnson (*Thought* 156-7) says that Digges' data on the New Star are "surprisingly" accurate. Digges dedicated *Alae seu* to Burghley and published it in late February 1573 and, although Tycho's *De Stella Nova* was in press before April 16, 1573, it was not yet printed on 3 May, 1573. Thoren (69) marvels at Tycho's reluctance to publish and, since Digges' work became available before Tycho's, he suggests that, by May, 1573, Tycho "had seen some other writings on the star." Perhaps, these were the data tabled prominently at the beginning of *Alae seu*.

In 1602, Kepler published Tycho's *Progymnasmata* wherein Tycho devotes over thirty pages to Digges' work, nearly twice the space allotted to any other on the subject. With much ado, Tycho shows that an obvious misprint in *Alae seu* leads to a silly result, whereupon he concludes, gravely, that Digges' data are close to his own. Tycho can ill afford to write sarcastically because his own data are not above reproach. Moreover, his measurements of the position of SN 1572 are the only ones of his entire career that have been lost. They alone "were not copied when one of Tycho's assistants later compiled a notebook that constitutes the sole record of his observations from his first efforts in 1563 up to those in December 1577" (Thoren 55). Given Shakespeare's persistent reference to Diggesian treatises and the commonality of Digges' and Shakespeare's dogged derogation of all things scholastic, the likelihood that Shakespeare intended the "star that's westward from the Pole" to serve as a tribute only to Tycho, is slim.

Holding Court.

In 1.2, King Claudius holds court as if to the manner born. The first item of business concerns Young Fortinbras, who seeks to recover lands forfeited on the death of his father. Claudius dispatches Cornelius and Voltemand to seek the intervention of Fortinbras' ailing uncle, Norway. He then grants Laertes leave to return to France. Shakespeare does not divulge Laertes' precise destination until 2.1 when Polonius instructs Reynaldo on how to spy on him. The subtlety with which Shakespeare slips in the connection to Paris belies the city's allegorical significance, as we shall see.

The king next turns his attention to Hamlet and addresses him as his "son." In response, Hamlet mutters, "A little more than kin and less than kind." "Less than kind" could mean "less than kindred," indicating that, despite appearances of kindness and kinship, Claudius does not have Hamlet's best interests at heart. Claudius asks why Hamlet is still so dejected at the death of his father, "How is it the clouds still hang on you?" and Hamlet responds, "Not so my Lord. I am too much in the sun." When Hamlet puns on "son" and "sun," he announces that he is devoted to his father and, in an elective monarchy, considers himself a legitimate, even favored, contender to the throne. Hamlet does not say simply that he is "in the sun," which would give the impression of his basking in the beams of Sol while enjoying a favored status, but that he is *too much* in the sun as if he is about to get burned.

Inky Cloak.

Hamlet insists that he has within him that which passes show and that his inky cloak and suits of solemn black are but the trappings and the suits of woe. Shakespeare frequently uses clothes as a metaphor to conceal the naked truth, as when Hamlet says:

124

> Seems, madam? nay, it is, I know not seems.
> 'Tis not alone my inky cloak, good mother,
> Nor customary suits of solemn black ...
> That can denote me truly.

Hamlet's "seems" are the "seams" of his "inky cloak." Gertrude's advice, "Good Hamlet cast thy nighted colour off," is also the playgoer's challenge, which is to peel off the layers of obfuscation to uncover underlying meaning. Just as celestial appearances do not necessarily represent physical reality, so we must cast off habiliments that shroud the sub-text. We do not judge a book by its cover, nor Hamlet by his appearance.

The king puns on Hamlet's "mourning duties." He refers to Hamlet's sadness at the loss of his father and, unknowingly, to the promise Hamlet will make to his father's spirit in the wee hours of the morning. Claudius upbraids Hamlet for continuing to mourn the loss of his father. The king says that the time for obsequies is over:

> But you must know, your father lost a father,
> That father lost, lost his, and the survivor bound
> In filial obligation for some term
> To do obsequious sorrow.

Claudius refers to past geokineticists and atomists, all of whom strove to establish their models but failed and passed on. He thinks that Hamlet's woe is unprevailing because geokineticism is a lost cause, and he wishes that Hamlet would learn from past mistakes. The implication is that Young Hamlet holds Old Hamlet in high esteem and thus, implicitly, upholds a view that's "incorrect to heaven." In the next line, Claudius says that Hamlet's heart is unfortified and his mind impatient, which comes "close to pronouncing Hamlet's behavior as heretical" (Sohmer *Mystery* 232). Claudius regards Hamlet as simple and unschooled because schoolmen do not tutor him.

John Dee was England's foremost mathematician and astronomer and he tutored Thomas Digges ably as, no doubt, did Thomas' father. From his publications and his later reputation as a scientist, mathematician, and engineer, Thomas was, in reality, well schooled, as is his alter ego, Hamlet. Rather, Claudius is the ignorant one as manifest by his support of the naive and popular interpretation of celestial phenomena, "For what we know must be, and is as common / As any the most vulgar thing to sense." Little does Claudius realize that it is not in his interests to restore Hamlet's sunny disposition.

Retrogradation.

Claudius expresses the royal "opposition" to Hamlet's desire to resume learning. He tells Hamlet that his intent "In going back to school in Wittenberg" is "most retrograde" to the desires of the royal couple. In the sixteenth century, many Danes matriculated to Wittenberg University, so it is not surprising to find that Hamlet is a student there. Hamlet succumbs to the pressure and agrees to remain at Elsinore, which pleases the royal couple, but their victory is Pyrrhic for they will sustain great losses.

Prior to speaking of retrograde motion, Claudius said, "Why should we in our peevish opposition / Take it to heart?" "Wittenberg," "retrograde," and "opposition" occur in the same literary and astronomical context. For Superior Planets like Mars, retrograde motion occurs at the time of Opposition (see Figures 1 and 4), and Wittenberg is renowned as the home of the first school of Sun-centered cosmology where the solution to the puzzling phenomenon of retrograde motion was first taught.

The astronomical use of the terms "retrograde" and "opposition" dates back to Chaucer in the fourteenth century. "Retrograde" also means "opposed, contrary, or repugnant to something," "tending or inclined to go back to an inferior or less flourishing condition" and, in the more literal sense of the word in use by 1564, "moving backward" or "returning upon a previous course" (*OED*). Hamlet learned of heliocentricism at Wittenberg, but his new parental configuration wants to keep him within the confines of geocentric conformity and does not condone retrograde motion to the site of heliocentric subversiveness. Shakespeare's only other use of the word "retrograde" is in an exchange between Helena and Parolles that concerns Mars, the Superior Planet with the most pronounced retrograde loops. The humor of that passage is that the warrior Parolles often retreats, which reveals the double meaning and suggests similar usage in *Hamlet*.

Retrograde motion contradicted the simple and theologically satisfying postulate of uniform circular motion. It cast doubt upon the simplicity of the celestial clockwork and raised questions about the manner of operation of the *Primum Mobile*. Claudius blames nature for spoiling what would otherwise be a beautiful and harmonious theory of the Universe. "Fie, 'tis a fault to heaven ... a fault to nature," he says. He thinks that retrograde motion must arise from an error committed by the Creator. In Claudius' mind, departure from the ideal forms that ancient philosophers envision, is nature's fault. Evidently, the Creator could not be bothered with detail as minor as keeping the planets moving steadily in one direction relative to the stars. The Creator committed other errors, too. Schoolmen debated why solid water floats in liquid water instead of sinking as solids should; and SN 1572 and SN 1604 were regarded once as matters of little consequence and a similar eruption observed by Hipparchus in the second century BC "was far too long ago to give serious cause for anxiety" (Coffin *Philosophy* 124). Obviously, the Creator did not intend for natural manifestations of ideal forms to be perfect always. Claudius' confirmation bias is so strong that he blames nature for phenomena that do not conform to his own flawed thinking. Thus, by opposing Hamlet's return to Wittenberg, Claudius opposes heliocentricism and identifies himself with the geostatic theory of Claudius Ptolemy.

In 4.7, Claudius describes his relationship to the queen as "conjunctive." The *OED* uses this example to illustrate that "conjunctive" can mean "having a relation of conjunction or union." Thus, "conjunctive" refers to the social and political union of Claudius and Gertrude. The earliest form of "conjunction" was that used by Chaucer in about 1374, to mean the action of conjoining in a common purpose, and the astronomical meaning emerged shortly thereafter. The two meanings are the first recorded usages and it seems plausible that Shakespeare intended both simultaneously. Only 14 lines separate "retrograde" and "opposition," whereas 15 scenes separate those two terms and "conjunction," yet all concern planetary alignments. A possible reason is that "retrograde" and "opposition" relate closely to Claudius and Hamlet in whose metaphysical roles retrograde motion and opposition figure prominently, whereas the queen's role is literally little more than that of mother and wife. Gertrude's conjunction plays a lesser role as she is a real-world source of support to her new husband and serves no sub-textual purpose (see Tables 2 and 3).

Dram of Eale.

In 1.4, in his "dram of eale" speech, Hamlet meditates on the circumstances of his birth. Sohmer ("Note") has noted that a dram of eale is a portion of old liquor added to new stock to establish its enological heritage, just as a father begets and serves as a role model to his son. Sohmer calculates that Hamlet was born 53 days before his parents' wedding and notes that, in those days, "vicious" meant "immoral" or "bad."

Perhaps, the "vicious mole of nature" in Hamlet is Old Hamlet's immoral imprint of the sin of fornication and bastardy upon his son. In the sixteenth century, society did not particularly frown upon such transgressions, but Shakespeare may have another goal in mind. Forasmuch as Hamlet personifies Thomas Digges, we wonder whether the author of the New Astronomy is illegitimate, in which case we would expect Leonard and wife to do everything they could to inoculate their child against the stigma. One way is purposefully to forget the child's year of birth. Thomas Digges' birth date was neither chiseled on his tomb nor otherwise recorded and, when sources do venture to state his year of birth, most often they cite the value simply as *circa* 1546. This is the known year of birth of Tycho Brahe, who is the astronomer whom the courtiers personify. Shakespeare confirms the proximity in time when he says that the courtiers are "neighboured" to Hamlet.

With the mooted identifications of Tables 1 and 2, the Bard would have known that Thomas Digges had children and that his family showed signs of enduring for generations to come. It is comforting to know that the grace of forgiveness enables Leonard to serve time in sulfurous flames and wipe his slate clean so that the soul of his son, in its time, can rest peacefully. From 5.2, we learn that Hamlet's soul does go to Heaven, implying that Thomas inherited no sin from his father. Thus, for example, Leonard's great-grandson, Edward Digges (1621-1676), Governor of the Virginia Colony from 1655 to 1667, could cultivate his lands and promote the silk trade free of ancestral stigma.

In the sixteenth century, Denmark was an elective monarchy, becoming a hereditary one in 1660. In Denmark at the turn of the seventeenth century, primogeniture was not the case, but a monarch's son would have a better than even chance to succeed his father. Perhaps the noble electors by-passed Old Hamlet's son because of that vicious mole (Sohmer "Certain").

Rotters.

In 1.4, at midnight, a flourish of trumpets and firing of ordnance announce the start of royal revelry. Gunfire has satanic connotations and, shortly thereafter, Hamlet sees the Ghost for the first time. Horatio worries that it imperils Hamlet when it beckons him to follow it. Marcellus worries that "something is rotten in the state of Denmark." The specter must soon return to Purgatory to finish its sentence and, in the meantime, it busies itself by telling Hamlet a thing or two. In 1.5, it confirms that it is the spirit of Old Hamlet, tells Hamlet that Claudius murdered his father and enlists Hamlet to exact revenge upon Claudius. Hamlet, Fortinbras, and Laertes all seek to avenge their fathers' deaths, as does Pyrrhus in the poem recited by Player I.

Shakespeare depicts Claudius as able and deserving of respect, yet also as villainous. Some believe that Ptolemy borrowed intellectual property without attribution and that he may even have manufactured data. Ptolemy's erroneous or fraudulent observations were chiefly responsible for the perpetuation of the error of Trepidation (Berry sects. 58, 83). Comparing one who may have appropriated data to a murderer may seem like overkill, but note that real-life events portrayed on stage and the allegorical sub-text follow different sets of rules because they serve different ends and pertain to wholly different spheres of interest. Claudius' World is one of appearances, and his world is one of polite seeming that can smile and smile and be villainous. Claudius believes that he the kingpin of the Old Astronomy and, as such, is cosmically geocentric and anagrammatically egocentric.

The apparition speaks to a number of different states, such as life and death, peace and war, innocence and guilt, not to mention the state of politics in Denmark. The Ghost's interest in terrestrial affairs concerns Old Hamlet's murder, thrice characterized as "foul," but the identification of the Ghost with the spirit of Old Hamlet solves only one aspect of its provenance. A cosmic sub-text questions why the diaphanous epiphany directs Young Hamlet to avenge a murder that it deems "unnatural." From the fifteenth century "unnatural" means "not in accordance or conformity with the physical nature of persons or animals," and "not in accordance or agreement with the usual course of nature" (*OED*). Humans are rare among mammals in their willingness to kill conspecifics, but many deplore the practice and, on this account, murder merits the term "unnatural." However, something "supernatural" is also unnatural in the sense that it does not accord with the usual course of nature, so, by calling Old Hamlet's murder "unnatural," Shakespeare could also mean that it serves a supernatural purpose.

Things in Heaven and Earth.

The Ghost tells Hamlet that it must not divulge secrets of its prison house but could relate a hair-raising tale that would "Make thy two eyes like stars start from their spheres." It goes on to say that the promulgation of what belongs to the eternal world would make anyone's hair stand on end like the quills of a fretful porcupine. Some believe that the word "stars" above refers to planets falling away from the spheres that supposedly bear them. Oberon uses the same conceit, but the use is puzzling because there are no known instances of Ancient Planets falling out of their orbits. Perhaps "stars" refer to two supernovae known in Western Europe in the second millennium, SN 1572 and SN 1006, the latter seen from Switzerland and points south. Perhaps "stars" refers to starry streaks in the night sky commonly called "shooting stars." Having passed from natural to supernatural space, the Ghost would have learned of all these phenomena and the workings of the heavens but, of course, must not divulge any information because, if it did, the struggle for knowledge of the Cosmos would end.

A mere 150 lines after the Ghost speaks of the stellar darters, Hamlet tells Horatio that there are "more things in heaven and earth" than "are dreamt of in [y]our philosophy." "Philosophy" is "intellectual investigation, science," and "[y]our" indicates textual ambiguity in naming either Hamlet or Horatio (Edwards 1.5.167n). If "your" is favored over "our," Hamlet is probably telling Horatio that heliocentricism and

Tycho's work on the New Star are not the whole story behind the New Philosophy. Horatio is destined to survive to tell Hamlet's story and initially, although he knows the culture of Wittenberg, he is in the dark about the significance of the Ghost. By 5.2, Horatio will know the whole story of the New Philosophy and will relay it to future pioneers.

The temperature drops below the dew point and the Ghost senses that dawn approaches. It bids farewell, "Adieu, adieu, adieu. Remember me." In Q2, "Adieu" is written as "Adew" and "a dew," which together denote the state of leave-taking and a vaporous state (Andrews 1.2.130n).

Moles.

Hamlet is shocked at the Ghost's revelations. After Horatio and Marcellus catch up with him, Hamlet swears them to secrecy. By now, the penitent specter is dilly-dallying in the "cellarage" and, while still within earshot, it encourages the conjuration by commanding "Swear," which it repeats three more times. After the third time, Hamlet says, "Well said old mole, canst work i'th'earth so fast? / A worthy pioneer." The mythological significance of the number four and the four-fold exhortation suggest that events are occurring for the good.

Hamlet calls his father's spirit an "old mole" because Old Hamlet was a "pioneer," who was "a soldier responsible for excavations and tunneling" (Edwards 1.5.163n). Of course, the Diggeses are "diggers" of sorts because the pioneers undermine the foundation of geocentricism that is now entrenched at Elsinore. Shortly after 1571, when Thomas Digges reported the death of his father, a passage by Ludwig Lavater (1527-1586) appeared (Wilson *Happens* 81), "Pioneers or diggers for mettal, do affirme, that in many mines, there apeare straunge shapes and sprites, who are apparelled like unto other laborers in the pit. These wander vp and down in caues and vunderminings." In the same vein, the Spirit of Old Hamlet has come down from on high and has holed up below, like a mole in the subterranean chambers of its habitat. Shakespeare may capitalize on the coincidence that Thomas Digges is associated directly with a "mole" since, in 1581, in *Plan for the Improvement of the Haven and Mole of Dover,* Thomas described his work on this cinque port.

The identification of the subterranean mole as the spirit of Thomas Digges' father brings to light an intriguing timing of events. In 1572, when the New Star appeared, Leonard Digges had just died. The inventor of the perspective glass and facilitator of so many astronomical discoveries had passed away just before one of the most remarkable celestial events of the century. The Bard compensates for the bad luck by allowing Leonard's spirit to return to the place where he once worked.

Shakespeare uses the phrase, "in the earth," twice within five lines to affirm that the old mole works there, and again when Hamlet tells Horatio that there are more things "in heaven and earth" than he knows in his philosophy. With all the concern for Elsinore, one would think that matters and events *on* Earth might better serve the interests of the play, but the spirit of the chief pioneer is digging tunnels and chambers *in* the Earth, where he subverts the prevalent philosophy of Elsinore.

Earthly matters crop up repeatedly. Marcellus and Horatio identify themselves as "friends to this ground," as they truly are, given their sub-textual roles as pre-scientific purveyors of grounded knowledge. The Bard stresses the role of epistemology in the last lines of 2.2, where he lays so much of the allegorical groundwork. "I'll have grounds / More relative than this," says Hamlet. The related pioneers are landed gentry working the fields of the New Philosophy. The pun anticipates the heroic search for proof of wrongdoing that is better grounded than the mere say-so of a shady spirit. At the same time, Hamlet prepares the way for the pending cosmic synthesis that requires capturing a patch of dirt so important that Fortinbras and his army must trek all the way to Poland to reach it.

Therapists.

As the first act ends, Hamlet vows to don the cloak of madness. "As I perchance hereafter shall think meet / To put an antic disposition on," he says, and proceeds to reassure the Ghost, "Rest, rest, perturbed spirit." Committed to action, Hamlet laments his fate, "The time is out of joint: O cursed spite, / That ever I was born to set it right." He must "restore the disjointed frame of things to its true shape" (Edwards 45). By "frame" Digges means Universe, and Hamlet's allegorical role is to straighten it out.

In 2.2, Claudius tells the two courtiers that he cannot imagine what ails Hamlet, apart from the obvious fact of his father's death. Claudius is so imbued with the righteousness of his cause that he denies the depth of Hamlet's grief. The spies stick together like Siamese twins and accept their assignments in short speeches each with the same number of syllables. The royal couple mocks their complementarity, making them appear as halves of a single entity:

> Thanks Rosencrantz, and gentle Guildenstern.
> Thanks Guildenstern, and gentle Rosencrantz.

Shakespeare portrays the didymous functionaries as if they were a unit comprised of two parts because, together, they personify Tycho's geo-heliocentric model, which has two centers of motion. As in Tycho's model where Sun and Moon circle the Earth, so the twin sycophants are satellites of the geocentric king who has sought their help because the hybrid model is more beholden to geocentricism than to heliocentricism.

Metamorphoses.

After the spies arrive in 2.2, Claudius gets straight to the point and explains that he needs them to deal with Hamlet's "transformation:"

> Something have you heard
> Of Hamlet's transformation - so call it,
> Sith nor th'exterior nor the inward man
> Resembles that it was.

"Transformation" was used in the fifteenth century to mean, "changing in form, shape, or appearance" (*OED*). The scientific meaning is change of form without alteration of quantity or value occurring in accordance with a definite set of rules. In other words, substitution of a new set of coordinates involves transformation of the geometry by which one center converts to another. Remarkably, the first scientific use of the word "transformation" was in the sixteenth century by none other than Thomas Digges, in *Pantometria* of 1571 (*OED*). In addition, Shakespearean and Diggesian usage is connected because Hamlet's transformation draws attention to the fact that both Thomas Digges and his alter ego, Prince Hamlet, suffered a change in their state at the time of their fathers' deaths.

The inner transformation changes the wheels and rods of old Ptolemaic geometry and the motions they generate into the new Copernican geometry. The second part is exterior and refers to the Diggesian substitution of an outermost shell of Fixed Stars by a uniform distribution in space. Exposure to planetary truths at Wittenberg transforms Hamlet inwardly, and the Ghost's teachings transform him outwardly. Neither the planetary system nor the Firmament "resembles that it was." Hamlet's transformation is a two-step process that changes the Old Astronomy into the New, so it is easy to see why Claudius worries about Hamlet's state because a change in the origin of coordinates would transfigure the hierarchy of the Old Astronomy, dethroning and de-centering Claudius and imperiling both his monarchical state and the foundation of hierarchical medieval existence.

The idea of transformation from one state to another underlies Shakespeare's favorite source of mythology, Ovid's *Metamorphoses*. Hamlet's transformation lies at the heart of the mystery because it associates Hamlet's quest for political and cosmological change. The protean Hamlet is doubly transformed, one transformation serving the literal ends of the play and the other serving the allegorical. This conceit occurs also in *Spaccio de la Bestia Trionfante* of 1584, where Bruno writes of the close life-long bond of friendship between Fulke Greville and Philip Sidney (1554-1586) who were nurtured and brought up together and had the same sorts of interior and exterior perfections. In addition, Tycho attached copies of his portrait to books that he presented to prominent persons, so many of his portraits were in circulation around the time that Thomas Digges and Shakespeare might have laid hands on them. In particular, the *Super Ex Libris* portrait was accompanied by the inscription, "Here is Tycho Brahe's outer image to be seen; I wish that the Inner, the hidden, may shine more beautifully," a woeful plaint from one tragically disfigured. In describing the transformation of Hamlet's inner self, Claudius refers to the very type of transformation that Tycho hopes will portray him more beautifully.

Hamlet is a tale of metamorphoses, of transformations, of changes from one state to another, for people, government, and cosmic models. The play simultaneously reorders the Great Chain of Being and

134

redefines the nature of the Universe under the aegis of the Supernal Being. Gertrude speaks of her too much changed son. Ophelia becomes distraught over the two major blows she suffered in short order, the death of her father and the departure of Hamlet for England, and has transformation in mind as she wishes the royal couple well in their own transformation. Somehow, she senses that change will come even to so permanent an institution as the royal family.

Freedom.

Gertrude directs aides to escort Rosencrantz and Guildenstern to where Hamlet is. Hamlet tells them of his melancholia and complains that the bounded geocentric model has lost its appeal. The pair argues with Hamlet, who says that Denmark is one of the worst prisons in the world with many confines, wards and dungeons. Kronborg Castle was renowned for its dungeons in which, they say, prisoners were tortured. Tycho's castle had cells where he detained peasant debtors. Rosencrantz tells Hamlet that Denmark is too narrow for his mind, which prompts Hamlet's celebrated response, "O God, I could be bounded in a nutshell and count myself a king of infinite space, were it not that I have bad dreams."

"Infinite space" refers to Digges' vision of a firmament filled with self-luminous stars and contrasts with prisons that serve as metaphors for the bounded World models. To Hamlet, being a prince confined to Elsinore Castle is like being a serf imprisoned in a dungeon cell or a New Astronomer constrained by a bounded Universe.

"Bad dreams" refers to both the oppressiveness of Elsinore and the threat of persecution because, within a few lines, Hamlet says, "by my fay, I cannot reason," meaning that free inquiry about the Universe is proscribed. Polonius advocates imprisonment if Hamlet does not tell his mother what ails him. Rosencrantz threatens Hamlet's liberty, which Claudius acknowledges, "is full of threats to all." Digges' cosmology threatens pedantry and, allegorically, the false ruler of the Danish state. One supposes that Shakespeare, no less than other poets in the sixteenth century, would think twice about addressing the topic of unbounded space, yet, being a poet who by consensus was ahead of his time, he would wish to write of it nevertheless. Neither Hamlet nor his dramaturge can speak openly about revolutionary ideas for fear of repercussions, the mere anticipation of which would cause even the most redoubtable poet to have bad dreams.

"Nutshell" refers to the rigid shell of stars supposedly encasing all of creation in Tycho's model. A "nut" is a fruit with a hard shell and, from the fourteenth century, "a symbol of something of trifling value" (*OED*). In this sense, Tycho's *minutum* with its shell of stars was the most compact of the bounded models, and Ptolemy's, which was barely larger, both resemble nuts. Tycho's geocentric model also qualifies as a trifle of little worth because of its shaky pseudo-Copernican foundation. Loosely speaking, "nutshell" is to "infinite space" as the infinitesimally small bounded models of Ptolemy and Tycho are to the Diggesian infinity. The shell of the "nut" could refer also to frame enclosing Tycho in the portrait he sent to England, with its foundation planted on the Earth and its arch overhead symbolizing the vault of the stars. A nutshell could symbolize the cranium enclosing the mind of a geocentric cosmologist. The wordplay refers also to Nut, the ancient Egyptian sky goddess, arching over the heads of observers like a shell enclosing the Earth, a connection compatible with the fact that Tycho's model is a generalization of the so-called Egyptian model.

In the Copernican model, the distance of the outermost planet, Saturn, is about half the value of that in the early geocentric models, but it is unlikely that "nutshell" refers to the Copernican shell because Copernicus had to have an enormous empty space, his *immensum*, between Saturn and the stars. This does not resemble a nut in any way and Copernicus had the flexibility of mind to allow at least the possibility of an infinite Universe. Shakespeare's "nutshell" leaves little doubt that he is singling out the Ptolemaic and Tychonic models, and Rosencrantz's remark that Denmark is too narrow for Hamlet's mind acknowledges that Hamlet subscribes to neither. Horatio and Hamlet talk of the "mind's eye," which may refer to the visualization of the Diggesian model and the antinomy of stars that reach up without end and boggle the mind.

Anne of Denmark.

The passage on infinite space and Denmark's prisons appeared only in F1 of 1623. Editorial opinion is that these lines are a cut, not an insert;

i.e., their omission from the earlier Q2 is deliberate. Perhaps political considerations are to blame (Edwards 2.2.229-56n). Anne of Denmark (1574-1619) was the consort of James I (1566-1625) and might have taken offense at derogatory remarks about Denmark, even though, after her death when the passage re-emerged in F1, Anne's widower was still alive. Another reason, possibly the primary one, is that by 1623, it was safer to address the seeming infinity of the Universe of stars and its potential conflict with theology because, thirteen years earlier, Galileo had already announced the existence of stars fainter than the eye could see.

Melancholy.

Hamlet complains, "it goes so heavily with my disposition that this goodly frame, the earth, seems to me a sterile promontory; this most excellent canopy the air ... this brave o'erhanging firmament, this majestical roof fretted with golden fire ... appeareth no other thing to me but a foul and pestilent congregation of vapours." Edwards (2.2.280-90n) believes that Hamlet suffers from "world-weariness" and, certainly, he is weary of Elsinore's World.

In the sixteenth century, "promontory" meant a point of land that juts out, or anything that resembles it (*OED*), and the "Promontory of Noses" is where Tycho went for a prosthetic nose. The passage connects Tycho's disfigurement to the element Air and conjures up images of nasal stuffiness, of the oppressiveness of Elsinore and of the crowded quality of bounded space in Tycho's model. Tycho had two artificial noses that he secured with adhesive salve. The one that he used on important occasions was made of gold and silver blended to a flesh tone. The other was made of a lighter alloy of copper and other metals and was reserved for everyday use and for his burial. In directing Claudius where to seek the body of Polonius, Hamlet says, "you shall nose him as you go up the stairs into the lobby," but Shakespeare does not limit mention of Tycho's disfigurement to *Hamlet*. In *Troilus*, Cressida says, "I had as lief Helen's golden tongue had commended Troilus for a copper nose." By alluding to Digges' model in the context of air, Shakespeare refers also to the fanciful Pythagorean notion that space must be infinite for it to "breathe."

Edwards (2.2.280-90n) attributes Hamlet's complaints to a "campaign to mislead Rosencrantz and Guildenstern and keep them off the true scent." When Hamlet says that he does not know why he has lost all his mirth, he misleads the spies into believing that his worldview is a *result* of his melancholia, so they do not suspect that World views like their own are the *cause* of it. Within a few lines, Rosencrantz announces that the players are on their way, so that, when they do arrive, the courtiers are on the wrong track and do not suspect the real cause of Hamlet's disposition.

Centricity.

After Hamlet has disrupted the player' play and offended the king, matters come to such a pretty pass that, in 3.3, Claudius tells his courtiers that he plans to dispatch Hamlet to England in their charge. After reinforcing the monarch's beliefs in geocentricism and its divine status, Guildenstern warns Claudius that, with kingly centricity, comes a divine duty, "To keep those many many bodies safe / That live and feed upon your majesty." The monarch's life is "that spirit upon whose weal depends and rests / The lives of many."

The king's subjects depend on his "weal" or well-being. "Weal" may also be an ironical reference to Copernicus' use of the word "wheel" to describe what he perceives as majestic planetary revolutions. Rosencrantz warns that "the cess of majesty / Dies not alone" because geocentricists believe that the Earth occupies the center of the Cosmos just as the king is at the seat of power in the state. The system, he says, is a massy wheel:

> Fixed on the summit of the highest mount,
> To whose huge spokes ten thousand lesser things
> Are mortised and adjoined, which when it falls,
> Each small annexment, petty consequence,
> Attends the boisterous ruin.

To Aristotle, ten thousand was the ideal population limit of a political unit. Thus, if Claudius were to lose his position, all those 10,000 unfortunates would meet with disaster as well.

The sub-text suggests that the loss of the Earth's centricity also means the collapse of the geocentric Universe with its spokes, spindles and gears. Rosencrantz warns, "Never alone, / Did the king sigh, but with a general groan," because the king's ten thousand subjects could hardly survive the cascade of crystalline shards. Conditions are even worse for the Tychonic solution in which heaven is filled with orbs, each carrying a star, which are all nested, connected and compelled, somehow, to move in perfect unison. The Tychonic model also has supporting structures with the magical properties of both strength and invisibility, except that, in this case, there are many more pieces. If the Earth at the supposed center of these fantastic frames were to lose its central location, all these "petty" attachments would die. The falling stars and planets would provide entertainment by night, and those lucky enough to witness the ruin in daytime would see the falling Sun, truly a once-in-a-lifetime event. Rosencrantz and Guildenstern share traits with Chicken Little and Henny Penny, for whom the sky is also a limited geocentric structure, because the toadies and the fowl all fear a falling sky. No wonder they toddle off to warn their respective kings.

Rats!

In 3.4, Gertrude summons Hamlet to her chambers, for he has ruined the entertainment that was supposed to help restore his senses. Polonius listens from behind an arras. Hamlet mistakes him for a rat and kills him. Oops! "I took thee for thy better," he explains, implying either that Polonius ranks lower than a rat, or that Hamlet mistook Polonius for his "better," the king, in which case the king ranks on a par with a rat. It makes sense that the king belongs to the order Rodentia since Hamlet snared him in *The Mousetrap*.

Rats and mice are undesirable residents in the halls of civil discourse, whereas a mole, the spirit of Old Hamlet, goes about his business outdoors, sub rosa. Moles belong to the order Insectivora whose members consume insects, aerate the soil, and are somewhat beneficial to plants. Monarchs employ mole-takers because molehills are unsightly additions to their royal gardens. Claudius is himself a mole-taker for he has rid the world of one whose spirit is now a mole in the cellarage, and has set his sights on one who has a mole of nature in him. As Old Hamlet moves the earth, so Young Hamlet moves the Earth.

Commission.

Hamlet wishes his mother good night and drags Polonius' corpse away. Gertrude is all too human and is understandably upset. In 4.1, she reports the death to the king, who resolves to ship Hamlet to England forthwith. Claudius gives Rosencrantz and Guildenstern a sealed commission to deliver to the English instructing them to dispatch Hamlet "not to stay the grinding of the axe." Hamlet's head must roll, and the sooner the better. The spies warrant Hamlet's mistrust for, as he explains later to Horatio, he suspects that they carry a commission ordering his execution. He relates how his insight and writing skill, not to mention the diplomatic seal that he carried, enabled him to devise and secure an altered commission that named the tiresome twosome as candidates for execution, instead of him. By sealing this commission, Hamlet seals the fate of his erstwhile friends and the model they personify. This is Shakespeare's response to Tycho's request to Thomas Savile for poetical approbation. Tycho's proclivity for bombast did not endear him to the Bard and his vain letter has had a result opposite to the one intended. Hamlet is now free to resume pursuit of the king by whose surcease he will catch success.

Indifference.

After two days at sea, pirates abduct Hamlet and obligingly strand him on the Danish shore, while the guards sail on, oblivious to the sticky end that awaits them. Horatio receives a letter from Hamlet describing his escape. In 4.6, in an image from the art of artillery, he tells Horatio, "I have words to speak in thine ear will make thee dumb, yet are they much too light for the bore of the matter." His words are too light for the bore of cannon, another reference to the military interests of the Digges team. In 5.2, Hamlet tells Horatio how he feels about the deaths of the courtiers, "Why man, they did make love to this employment." Hamlet expresses indifference to their fate because they were a nuisance that he had to deal with before he could get on with his chief task and because, allegorically, they came between the mighty opposites of the Old and New Astronomy. Hamlet is unrepentant because the spies threatened his life, but he is allegorically indifferent because the model they personify was never a primetime player.

The Polish Plot.

In 2.2, Cornelius and Voltemand return from their mission to alert Old Norway that Fortinbras is demanding return of lands lost by his father. Old Norway discovered that Fortinbras gave the impression of wanting to attack Poland when, in reality, he was planning to attack Claudius. Senescent Norway created the military pretext for the pending cosmological synthesis because he forbade Fortinbras never more to take up arms against Claudius but to send his soldiers against "the Polack." The term supposedly refers to the King of Poland just as "the Dane" refers to the King of Denmark, but the Bard may have another Pole in mind as well. Fortinbras dispatches a Captain to secure permission to pass through Denmark en route to some part of Poland "to gain a little patch of ground / That hath in it no profit but the name."

The Norwegian sortie into Poland seems absurd unless seen in the present context. Blessing (160) writes, "The combined Danish - Norwegian - Polish - Carpathian - Transylvanian - Anatolian - Trans Caucasian - Persian - Afghan and Baluchistani forces under the supreme command of Fortinbras have reached the banks of the Indus River." The satire is poignant insofar as *scientiae* transcends human divisions and because all tribes are welcome in its tent. Science, like the Canon, is for everyone, and the nations listed in Table 1 are emblematic of its universality.

In F1, the interaction between Fortinbras and his Captain takes up only a few lines, which are important in preparation for the resolution of the literal plot, but in Q2, the exscinded passage helps a great deal to establish the cosmic context and set the stage for resolution of the allegorical plot.

The patch of ground is both famous and unsuited for agriculture. "I would not farm it," says the Captain, yet its capture apparently warrants a full-fledged military assault to "gain" it. Saxo states that graveyards are not arable, suggesting that the patch in question is a graveyard as well. This particular one must have great significance because of the "name" associated with it, suggesting, in the present context, that the site contains the remains of Copernicus. Shakespeare hardly refers to the King of Poland when he says that "the Polack" never will defend the ground. Rather, after 1543, Copernicus could not defend himself in any tribunal of this world against charges of heliocentricism.

Many have sought the location of the grave of Copernicus and it does seem odd that so prominent an intellectual and a church official in relatively high standing would have an unmarked grave. Perhaps Copernicus was already unpopular at the time of his death. Fortinbras needs an army of thousands to capture it, and the plot in question "is not big enough to hold those who are to fight for it, or to bury those who are killed." In 4.4, we learn that 2,000 souls and 20,000 ducats are "not enough to fight out the dispute" but 35 lines later, the army has swelled to 20,000 men. If these discrepancies are not textual errors, they indicate that, regardless of the precise number of soldiers involved, the graveyard is too small to hold many corpses. The military host is large, but Shakespeare imperils it hyperbolically as a measure of the opposition it would encounter if it were to restore value to the mortal remains of Copernicus.

Why is the cut in F1 so drastic? The answer may lie in events that transpired in the interim between Q2 and F1. As noted in Chapter 1, Tycho Brahe's data were essential to Kepler's discovery in the first two decades of the seventeenth century of the three empirical relations for planets. Of these, the third is the so-called Harmonic Law, whose nature conforms to Pythagorean and neo-Platonist philosophy. Since much of Q2 is devoted to belittling Tychonic contributions, perhaps the F1 editors saw fit to reduce the cosmic emphasis in general. A similar argument pertains to the accomplishments of Thomas Harriot (1560-1621), as the next chapter shows. In addition, Galileo's celestial observations of 1610 and the subsequent fallout led to increasing interference by ecclesiastical authority. At the same time, the Thirty Years War saw states like England and the Scandinavian countries pitted against the House of Hapsburg and Holy Roman Empire. The editors of F1, fearing that the purpose of the Norwegian plot against Poland might be as transparent as the plot in Poland is small, may have decided that it was in the national interest to omit lines that describe the allegory with such comparative clarity.

Constant vigilance is the right approach when navigating troubled waters and the F1 editors were prudent to exercise caution. By omitting 58 out of 66 lines in 4.4, the F1 compilers could have their cake and eat it, because some readers would surely wonder why Hamlet's fourth soliloquy was omitted. If challenged, the F1 editors could plausibly deny heresy but scholars with access to Q2 would raise questions that would reveal Shakespeare's thinking. Q2 suggests that the cosmic puppeteers have a schedule of their own and, as Fortinbras forges ahead with his mission, Hamlet prepares to screw his courage to the sticking-place. "Oh from this time forth, / My thoughts be bloody or be nothing worth," he says.

More Madness.

In 4.5, Horatio urges the queen to allow Ophelia to speak with her, lest Ophelia strew "dangerous conjectures" in minds intent on mischief. Ophelia enters, singing and distracted. One song concerns her father and a certain young man who promised to marry her before he came to her bed. She seems distraught and her babble switches back and forth between allusions to her father and Hamlet (Edwards 4.5.46-7n). Claudius enters and listens to her singing, prompting him to ask how long she has been in that state. Ophelia stops singing and switches to prose. She says, "I hope all will be well. We must be patient," and goes on to lament her father's death and his burial in the cold ground. "My brother shall know of it," she vows. No sooner said than done. Claudius announces that Laertes has returned in secret from France and that his presence is as serious as all of Ophelia's misfortunes combined (Edwards 4.5.86n).

Laertes arrives at the head of a throng that proclaims his candidacy for king. Seneca observed that the applause of a mob is proof of a bad cause and, indeed, in this case, the "rabble" does him no good. He bursts in upon Claudius and demands to know who killed his father. Claudius denies involvement. Ophelia re-enters, singing and lamenting, and Laertes calls her performance a "document in madness." After Ophelia leaves, Laertes raises the question of the stealthiness of his father's funeral. Claudius replies that "where th'offence is, let the great axe fall." Claudius is still of a mind to decapitate Hamlet.

Horatio receives a letter from Hamlet announcing his return to Denmark, as do the king and queen. Since Claudius no longer has his chamberlain and courtiers to lean on, he turns to Laertes and, in 4.7, explains that Hamlet conspired against him as well. Laertes makes the fateful decision to be "ruled" by Claudius and, in so doing, becomes an accessory to bounded geocentricism and fair game for Hamlet.

In the players' play, Hamlet baited the king and now the king seeks to return the favor. He inveigles Laertes to engage Hamlet in a swordfight. "Revenge should have no bounds," he says, meaning that he will stop at nothing to kill Hamlet. Claudius imputes revenge to the space of the deity, which has no bound. Laertes agrees to the fight because he seeks to avenge his father's death and the two conspire to envenom Laertes' rapier and to add lethal toxins to a drink for the nonce just in case Hamlet escapes the poisoned stick.

Gertrude announces that Ophelia has died by drowning. Her death saddens Gertrude as it does the literalists in the audience, but some suspect that the New Philosophy will not triumph through a liaison between Hamlet and Ophelia and they hope for a happy outcome in some other way.

More Pioneering.

In 5.1, a pair of Clowns prepare Ophelia's grave and one calls the other "goodman delver," or "master digger" (Edwards 5.1.12n). Julia Usher has pointed out that the pun is situational, for the delver digs into inner space to reach the truth about outer space. The other clown says, "The Scripture says Adam digged. Could he dig without arms?" "Adam digged" may be a reference to Thomas Digges' forebear, whose first name was Adomarus. The identification is likely since Adomarus was a judge under Edward II (1284-1327), and the reference to "Adam's profession" occurs immediately after the two clowns discuss points of law. Hamlet refers to "My father's spirit, in arms!" and debates "whether 'tis nobler ... to suffer ... slings and arrows ... or to take arms against a sea of troubles ..." Such technical words occur in *Stratioticos* where Thomas Digges refers to "men at armes." If this identification is correct, then his family receives mention in the same way as Tycho's, for both have ancestors who figure in the play.

As the diggers throw up skulls, Hamlet comments, "Here's fine revolution, and we had the trick to see't." Hamlet's remark begins with the present tense and ends with the past tense, suggesting that something changes between the "trick" and the "revolution." The primary meaning of "trick" dates to the fifteenth century and means "a crafty or fraudulent device ... an artifice to deceive or cheat." It is a clever device, a "contrivance or invention," as in *The Taming of the Shrew* where Shakespeare writes, "A knacke, a toy, a tricke, a babies cap." "Trick" can also mean a "characteristic quality" or "distinguishing trait," so the quote above implies also that the Diggeses possessed the distinguishing traits to see the Universe in a new way. Thus, in a parody of the Magi and their alleged deceptions, Shakespeare refers to the forerunner of the telescope, which many believed was a fraudulent device. By 1576, Leonard Digges was dead, yet even from the grave he is instrumental in overturning the Old Astronomy. The *OED* cites this very passage to explain "revolution" as "alteration, change, mutation." Its astronomical meaning (the orbital motion of the Ancient Planets) was in use by 1390. By 1450, the word came to mean "great change or alteration in affairs or in some particular thing." When in 1543 Copernicus made the word "revolution" essentially the entire title of *De Revolutionibus*, the possibility of a double meaning was already in place and, if Copernicus did not intend a double meaning, Shakespeare surely would have.

There is more to the plot than the grave of Copernicus, however. The epithet "subversive" applies equally to these diggers because the word derives from the Latin *subvertere*, which means literally "to turn from beneath," like radicals who plant seeds of change that germinate, take root and transform perceptions. The chief digger is a "Clown" because of the lowliness of his profession, and Leonard Digges is the old mole that is a low-down digger as well. Shakespeare celebrates the accomplishments of diggers of celestial data but makes it known that they, too, have feet of clay and, eventually, will need the services of the very profession that their name resembles. In *Hamlet,* all are laid to rest in a rather peremptory fashion (Sohmer *Mystery* 232), the lesson being that we are all equal when we are six feet under. Shakespeare is an equal-opportunity satirist because *he who [is] "digs"* is as much of a Clown as anyone is.

In the midst of all the puns, Hamlet says, "We must speak by the card or equivocation will undo us." Shakespeare must scatter an abundance of clues lest the cosmic implications of *Hamlet* pass unnoticed, but, at the same time, he must disguise meaning sufficiently well that the cosmic expose not imperil England – or him.

The gravedigger identifies the young Hamlet as the one sent to England to recover from madness. Allegations of madness afflict two other late sixteenth-century scholars who advocated a fresh look at the World. Tycho was thought mad when he fell into disfavor with the Danish court, and Robert Greene (1558?-1592) accused Marlowe of blaspheming with the mad priest of the sun, Giordano Bruno.

Shakespeare goes to great lengths to establish that Prince Hamlet is 30 years old when he kills the two personifying Tychonists and is about to finish off Claudius. In 1576, Thomas was about thirty when his *A Perfit Description* signaled the demise of the Old Astronomy. In the same year, the Danish Parliament passed a law making it illegal for a man who murdered his brother to inherit the victim's estate and, at one time, the Digges' clan itself came close to fratricide. If the death of Leonard Digges and/or the advent of the New Star date Act 1 of *Hamlet* to 1572, and Digges' *A Perfit Description* dates Act 5 to 1576, then the coincidence exists that the number of acts of *Hamlet* equals the number of years from 1572 to 1576 inclusive.

Many wonder whether Hamlet could still be a student at the ripe old age of 30, arguments being that his classmates at Wittenberg would be more than ten years younger and surely no one would remain a student that long. However, Copernicus was about 33 when he returned to Frauenburg in 1505 or 1506, Rheticus was 27 in 1541 when he returned to Wittenberg after studying under Copernicus, and Thomas Digges was 30 when his *Perfit Description* went on sale. Statistics begin with sample sizes of three, so we may conclude that, in the sixteenth century, an age of 30 is about average for students at the cutting edge of the New Astronomy.

Wager.

In 5.2, Hamlet expresses regret at his treatment of Laertes and, as if on cue, Osric arrives at the king's bidding to entice Hamlet into a swordfight with Laertes. A prolonged dialogue ensues that has long baffled experts (see Chapter 9). Equally puzzling are the terms of the king's wager, which Osric presents, "The king sir hath wagered with him [i.e., Laertes] six Barbary horses, against the which he has impawned, as I take it, six French rapiers and poniards, with their assigns, as girdle, hangers, and so. Three of the carriages in faith are very dear to fancy, very responsive to the hilts, most delicate carriages, and of very liberal conceit." Poniards and assigns are daggers and accessories; a girdle is a sword-belt, and hangers are straps that hold swords. Osric deigns to mention the rest of the accoutrements, dismissing them, as if with a wave of the hand, with the words "and so [on]."

Hamlet needs clarification and asks, "What call you the carriages?" A carriage can mean the transport of a cannon or, as Osric explains, a hanger. The *OED* cites no other instance where "carriage" means "hanger," suggesting that the word is affected, like Osric's description that they are "very dear to fancy," "very responsive to the hilts," "most delicate" and "very liberal." Hamlet accepts Osric's use of the term "carriage" by saying, "I would it might be hangers till then." Evidently, these three carriages are special and not like those run-of-the-mill hangers that you find everywhere.

Hamlet sums up the terms, "six Barbary horses against six French swords, their assigns, and three liberal-conceited carriages – that's the French bet against the Danish." The Danish bet must refer to the king's bet, so the "French" bet must be that of Laertes. It is true that Laertes has spent time in Paris and that French rapiers are part of his wager, but is not Laertes still Danish?

Odds.

Osric continues, "The king ... hath laid ... that in a dozen passes between yourself and him [i.e., Laertes] he shall not exceed you three hits." Many assume that a pass and a hit are the same and that a bout or round ends after each hit. They take this description to mean that Laertes wins as soon as he makes eight hits, and that Hamlet wins with only five hits, for a handicap of three hits; but the terms lack precision and no one has adequately explained them. In the standard interpretation, if Laertes wins eight rounds, Hamlet may have won anywhere from none to four, but cannot have won five because the contest lasts only twelve rounds. Similarly, Hamlet wins with five hits because Laertes cannot then win eight. By these rules, a tie at six hits is impossible because Hamlet would already have won with five hits.

Hamlet accepts the terms of the wager and tells Osric that he will try to win for the king but that if he does not, he "will gain nothing but my shame and the odd hits." The *OED* cites this passage to exemplify the meaning of "odd" as "extra," or "given over and above," as in, "You shall have 40 shillings and an odd bottle of wine." The same meaning occurs in the phrase "odds and ends." If Hamlet loses, he will gain "the odd hits," which are hits that score in his favor but are "odd" because they are "given over and above" the hits that really count, which are the ones that Laertes inflicts to win.

Hamlet says that he has been in continual practice ever since Laertes went to Paris and that he shall win at the odds. Hamlet acknowledges that the king has laid odds on the weaker side, but Claudius is not worried. He handicaps Laertes because, he explains, he has witnessed the swordsmanship of both and judges that Laertes' has improved. In addition, unbeknownst to the audience until now, a Frenchman, Lamord, had visited the king two months earlier and had praised Laertes' skill. This would have occurred about the time of Old Hamlet's death, which is another French connection potentially inimical to the Hamlet family.

Carriages.

Osric completes the terms of the engagement. "He hath laid on twelve for nine," he says, where "he" refers to Claudius. Coming right after the statement on the three-hit differential for a win, some believe that this twelve-to-nine ratio refers to bookie odds. Others believe it refers to the relative value of the stakes. Each option faces difficulties. The ratio 12/9 is the same as 4/3, but, for some reason, Shakespeare leaves it to the audience to divide out the common denominator, 3. The difference 12-9, happens to equal the assigned handicap of 3 (which is 8-5 in the interpretation mentioned above) but the sum, 12+9, does not equal the number of rounds allowed nor is it related to the problem in an obvious way. It is unlikely that Shakespeare means "twelve for nine" to refer to a probability in the modern sense since probabilities range from naught to one and $12/9 > 1$. Moreover, when Claudius speaks of "twelve for nine" he could surely not mean that the probability of a Hamlet win is $12/(12+9) = 12/21 = 4/7$, which is only slightly better than fifty-fifty and not much of an incentive for Hamlet to accept the terms. If the numbers 12 and 9 are the actual values of the respective wagers, Osric does not state the unit of currency. He would not need to if he means that 12 for 9 is only a ratio, but then, again, why does Shakespeare leave the arithmetic to us?

The problem reduces to reconciling "twelve for nine" with the two stakes. The king's stake is six Barbary horses and Laertes' stake is six French swords and three liberal-conceited carriages. These stakes qua numbers are 6 and 9 and do not compare to the stated integers 12 and 9, but 6 and 9 could transform to 12 and 9 by doubling 6. Is such a doubling justified?

The answer lies in Hamlet's query, "What call you the carriages?" and Horatio's immediate comment, "I knew you must be edified by the margent ere you had done." Somehow, Horatio knows that "the margent" will edify Hamlet, but "carriages" is plural and "margent" is singular, suggesting that Hamlet and Horatio are not talking about quite the same thing. The mystery unravels upon examining the meaning of "margent" and its more common equivalent "margin." Primarily, "margent" means the space between the edge of a page and the text. It is also used figuratively to mean the "edge" or "border" of something, as in Spenser's *Faerie Queene* of 1596, "From th' utmost brinke of the Americke shore Unto the margent of the Molucas" (*OED*), which refers to the Molucca Islands lying between the Philippines and Australia. The word "margin" can denote "an extremity or furthermost part of something" as in usage from about 1595, "eu'ry Margine of this earthy sphere." From the fifteenth century, a "margent" is the "ground immediately adjacent to a river or body of water ... a shore," as in the margin and the brink of the sea. Furthermore, as noted, things French figure prominently. Laertes visits France twice; there are French rapiers, a "French" bet, a Frenchman who values Claudius and speaks well of Laertes and, in exemplifying the use of "margent," the *OED* specifically mentions the sea past France. If that sea is the Mediterranean, the ground on the other margent of that sea is Africa – specifically, the Barbary Coast, home to the very breed of horse in the king's wager.

The sub-text provides the context for understanding the arithmetic. The king thinks he lives at the hub of creation, making Elsinore the origin of coordinates from which all distances on Earth are measured, just as, in a geocentric Universe, the Earth is the origin for universal distances. This is the very bias that Hamlet's transformation overcomes and about which the sub-text of *Hamlet* is concerned. Shakespeare parodies Claudius' regi-centricism just as he does Tycho's Ven-centricism by drawing attention to the fact that (in round numbers) the distance from Elsinore to the Barbary Coast – whence come the horses – is a factor of two greater than the distance from Elsinore to Paris – whence come the "French" items. With the help of the ratio of the distances to the Barbary margent, and to Paris, as measured from the center of Claudius' Universe, Shakespeare derives the factor of two that establishes the relative worth of the king's "Barbary" wager vis-a-vis the "French" wager.

Horatio comments on the "margent" shortly after Hamlet inquires about "carriages," which emphasizes the fact that the Barbary Coast lies at the margin of the narrow worldview espoused by the Danish monarch. Thus, the first number 6 that concerns the actual number of the king's wagered items, receives a weight of 2 and results in a relative worth of the items of 12 to 9. Claudius flatters and coerces Hamlet, not just by handicapping Laertes but also by wagering a greater amount on his winning. The matter of stakes is part of the yarn that the king spins to hide the fact that he has fixed the odds. The king expects Hamlet to die, if not by envenomed sword then by poisoned chalice, and deceit, not skill, will decide the outcome. In reality, Hamlet's life is at stake and not horses, swords, and carriages. As long as there is honor among the deceitful and the two swindlers regard the stakes as fictitious and irrelevant to their greater purpose, each stands to win by exacting revenge on Hamlet and neither suffers any material loss.

The Fight.

Just before the fight starts, the king announces that:

> If Hamlet give the first or second hit,
> Or quit in answer of the third exchange,
> Let all the battlements their ordnance fire.

Let H and L denote hits by Hamlet and Laertes, which we assume are independent of any previous outcome as long as wounds are not severe, but to those who know of the conspiracy to kill Hamlet, these lines express the king's secret hope that Hamlet will die within three rounds: either by suffering a wound in the first round (L), or if not then, in the second round (HL), or if not then, in the third round (HHL).

Ambiguity clouds the meaning of the command. In the first line, to "give" can mean to "inflict," but, from 1582-1593, it can also mean the opposite, to "yield" or to "concede," as in "I give thee a victorie," "to give no foot of ground" (*Henry VI part 3*) and "enemies ... driven to give place" (*OED*). In brief, to "give" a hit can mean to "inflict" or to "concede" a hit, which are mutually exclusive outcomes. After the first round, the score is H or L regardless of the meaning of "give," for if Hamlet "gives" a hit, either he inflicts it, H, or he receives it, meaning that Laertes inflicts it, L, or vice versa. In order for the cannoneers to fire their ordnance in accordance with the king's command, they would need to know which of these meanings obtains. In the second round, "If Hamlet give the first or second hit" means either that Hamlet *inflicts* the first or second hit so that the sequence HL or LH is the case, or that Hamlet *suffers* the first or second hit so that the sequence LH or HL occurs. Either way, the outcomes are the same, either HL or LH. The gunners' dilemma is moot because Hamlet inflicts both of the first two hits, HH, and the cannons remain mute.

The final option, "Or quit in answer of the third exchange," is ambiguous by virtue of the meaning of "quit," again putting the gunners into a quandary. The customary meaning of "quit" in this context is the general sense of playing one's part (*OED*), as when Hamlet strives to win each bout. Dating from the fifteenth century, however, the verb "quit" also means to "give up" or "cease to be engaged in, or occupied with," often used in the sense of "to leave," "to leave the premises," or "to depart from a place or person." The gunners do not know which meaning of "quit" applies. If Hamlet stops fighting because he has won three rounds (HHH), ordinance would fire as commanded, which would accord with the king's desire to save appearances and dupe the court into thinking that he champions Hamlet. If "quit," means that Hamlet expires owing to Laertes mortally wounding him (HHL), then ordinance will fire in celebration of Hamlet "quitting" the scene, feet first.

The contest begins, and Hamlet inflicts "a very palpable hit," *H*. This knocks out the king's first secret hope, *L*, because now Hamlet lives to fight a second round. Just in case, Claudius tries his luck with his back-up plan by offering the poisoned refreshment to Hamlet, who declines it. When Hamlet makes the next hit as well, *HH*, the king says, "Our son shall win." Hamlet is gaining on Laertes and Claudius is saving the appearance of an honest bettor by mouthing words that others would construe as gleeful. In reality, the king is alarmed because Hamlet has made the first two hits and, from what Lamord told him, Laertes is highly skilled and should have ended the contest already. Gertrude takes the king's remark literally and, in anticipation of her son's triumph, partakes of the viperous distilment. Sadly, the king is too slow to forestall her untimely celebration.

Laertes then wounds Hamlet, giving the sequence *HHL*. The cannoneers continue to hold their fire, however, because they know nothing of the king's deceit and because Hamlet is still standing and has not "quit." There is no time for the king to clarify his order because the plot thickens faster than he can articulate the good news. The contestants grapple and accidentally swap swords. The king notices this, becomes even more alarmed, and orders that the two be separated. His worst fears are realized when Hamlet wounds Laertes with the envenomed rapier.

The queen falls and in her dying breath tells of the poisoned chalice. Laertes, who has not "quit" the scene either, lays blame upon the king. Despite being at death's door, Hamlet wounds Claudius with the same rapier and forces him to drink from the poisoned cup, so the king suffers a condign death. Laertes tells Hamlet that he is mortally wounded and, having exchanged rapiers, the two exchange forgiveness as well. Laertes absolves Hamlet of guilt for the death of his father and proclaims his own innocence, "Mine and my father's death come not upon thee, / Nor thine on me." Laertes gives up the ghost, and Hamlet exculpates him for the error of his ways. "Heaven make thee free of it!" he exclaims. Hamlet prepares to step into the hereafter. "I follow thee," he tells the expired Laertes. Horatio proclaims, presumably on good authority, that Hamlet's soul is destined for Heaven, so both sons are free of the sins of their fathers.

Resolution of the allegorical plot begins with the arrival of Fortinbras from Poland, at which time the cannoneers finally have reason to fire their ordnance. Osric declares:

> Young Fortinbras, with conquest come from Poland,
> To the ambassadors of England gives
> This warlike volley.

This is the moment when Shakespeare connects the Copernican model from Poland to the Diggesian from England. The cannons fire in celebration of the birth of the New Philosophy and not of its demise. Throughout, Hamlet and Fortinbras have marched to the beat of a Distant Drummer and natural philosophy is now no longer solely the province of theorists.

In his dying voice, Hamlet names Fortinbras as the next king, "I do prophesy the election lights / On Fortinbras." Horatio has witnessed enough of the revolution and will have no difficulty filling in the details, so he agrees to inform the yet unknowing world. Such momentous events need recording for posterity, and Hamlet says that Horatio must live on to tell his story.

Martial Themes.

Fortinbras accords the deceased hero full military honors. The four pallbearers signify the natural order, which is the true model of the World for which Hamlet gave his life. In mythology, four was a proper numeral for Hermes-Mercury, companion to the Prince of Planets, Apollo, and mediator between Heaven and Earth. The words uttered by Fortinbras eulogize Thomas Digges in a manner appropriate to the military interests listed on Digges' tombstone. Hamlet would have given a good account of himself had he been put on, in agreement with the opinion that Thomas Digges would have proved himself in combat had circumstances, particularly his engineering and mathematical skills, not steered him into other service.

Fortinbras' praise signifies also that Digges will join the ranks of his predecessors who waged war on behalf of the New Astronomy. Under Fortinbras, free inquiry will enjoy the protection of the state and, in theory at least, future pioneers should no longer suffer bad dreams. The implication is that the power of the military will cement the triumph of the New Philosophy.

Hamlet has fought the good fight. He has finished the course. He has kept the faith. He has foregone the crown of Denmark and, henceforth, there is laid up for him a crown of righteousness in Heaven (cf. II *Tim.* 4:7-8). The death of the hero is a heavy price to pay, but *Hamlet* is a tragedy in appearance only because the hidden reality has a different cast. Shakespeare leaves the play's final word to the one who holds up a ray of hope for the future. In what could be a warning of future peril, Fortinbras orders the final tribute to the fallen hero, "Go bid the soldiers shoot." Hamlet's last words, "The rest is silence," may mean that, henceforth, no beneficent deity will step up to guide humanity in pursuit of knowledge of the natural world.

CHAPTER 9: FURTHER IDENTIFICATIONS

Great wits are sure to madness near alli'd,
And thin partitions do their bounds divide.

John Dryden

Shakespeare names Hamlet and Claudius for famous historical figures associated with the theme of *Hamlet*. Along with Rosencrantz and Guildenstern, they personify the chief cosmological models vying at the turn of the seventeenth century. The expectation is that other characters receive appropriate names as well. Table 2 summarizes results.

Polonius and Reynaldo.

As a dyed-in-the-wool foe of pedantry, Shakespeare could well have created Polonius as a caricature of Robert Pullen (d. 1147), a medieval schoolman and one of the founders of Oxford whose latinized name is *Polenius*. Reynaldo takes his name from John Reynolds (1549-1607), a contemporary of Shakespeare and President of Corpus Christi College, Oxford. Reynolds was an inveterate enemy of the theatre who, in 1599, published a diatribe, *Th'overthrow of stage plays*, expressing opposition to the staging of plays, and the juxtaposition of the names Polonius and Reynaldo in 2.1 is likely a slight upon these Oxford dons. The suggestion is plausible given the possibility that when *Hamlet* was performed at Oxford the names of the two characters may have been changed to avoid trouble for the players. Shakespeare would not have much time for either Pullen or Reynolds and the play-within-the-play serves also to express disapproval of Reynolds' position because it demonstrates the utility of stage plays in revealing truth.

Ophelia.

Saxo does not name the woman who tempts Amleth, and Shakespeare would feel free to select a name for the equivalent maiden in his tale. Shakespeare seems to have coined the name "Ophelia" out of whole

155

cloth (*OED*), perhaps from the Greek "to succor" or "to help." Perhaps, in tolerating her abusive father, she needs help herself. This explanation might satisfy the literal interpretation, but symbolism may underlie her dramatic relationships and her name, like those of other leading characters, might refer to an inherent feature of the cosmic allegory.

Shakespeare frequently uses the properties of deities of classical mythology and consistently associates rulers with the Sun, as do both Copernicus and Digges. Phoebus Apollo is god of the Sun, Diana is goddess of chastity and of the Moon and the Sun-chariot is the symbol of royalty. The Moon helps the Sun to rule heaven just as, during creation, God made "the greater light to rule the day, and the lesser light to rule the night" (*Gen.* 1:16). Hamlet has an amorous interest in Ophelia, who is thus, potentially, Hamlet's queen. Since Ophelia is supposedly a virgin as her brother's and father's concerns attest, and since Hamlet sees himself as filled with the light of the Sun, it seems likely that Shakespeare would invent a name for her that is associated with the Sun's companion, the chaste Moon.

The prefix *op-* is one of several variants of *ob-*, which has several meanings, one being "opposed to" or "facing." Another, often used with a coloring of the former, is "completely" (*OED*). In combination, these yield "completely facing." The *OED* lists combinations of *ob-* with words only of Latin origin, none of which begin with the letter "h." When combined with the Greek *helios* (the Sun), and using a feminine ending, we have *Op-heli-a*, denoting the Moon at Syzygy, i.e., at the time of Opposition or Conjunction, when one or other face is toward the Sun.

Shakespeare associates the Moon with water, as in the "moist star," the "watery star" from *The Winter's Tale,* and the "governess of floods" from *A Midsummer Night's Dream.* The Moon is the moist star because of her tidal influence on the "Neptune's salt wash," the oceans. The Moon symbolizes chastity, as when Oberon speaks of the "chaste beams of the watery moon." Spring tides are highest and occur when the Moon is at Syzygy. During high tide when Neptune's flood is deepest, prospective swimmers must sink or swim. Polonius warned Ophelia that if she were to develop a relationship with Hamlet she would be out of her depth. "Lord Hamlet is a prince out of thy star," he says, where Ophelia's "star" is, reasonably, the Moon.

Further evidence that the Bard named Ophelia for the moist star is found when, on hearing news of her drowning, Laertes exclaims, "Too much of water hast thou, poor Ophelia." Ophelia did not drown herself according to the established law according to which if "water comes to [her], and drown [her], [she] drowns not [herself]." In Ophelia's waning moments, she seems a bit touched, in keeping with Othello's description of the aberrant Moon:

> It is the very error of the moon;
> She comes more nearer Earth than she was wont,
> And makes men mad.

Ophelia flounders in the ebb and flow of events and, somehow, a body of water is to blame. The name "Ophelia" may have an association with Helen, who signifies either the Moon or a basket carrying offerings to the moon-goddess. In 4.5, when Ophelia enters distracted and carrying flowers in a basket, her actions and speech again reflect her lunar attributes.

The topic of seawater arises again in Hamlet's case when, in 1.4, the Ghost beckons Hamlet to follow it. Horatio fears for Hamlet's safety:

> What if it tempt you toward the flood my lord,
> Or to the dreadful summit of the cliff
> That beetles o'er his base into the sea?

Horatio worries needlessly. Hamlet survived his sailing adventure and is not at odds with Neptune. His problem is that he is too much in the Sun.

Barnardo.

Several notable candidates have names that resemble Barnardo's. Bernard de Chartres (d.*c.*1130) headed the cathedral School of Chartres in France and was the principal Platonist in Western Europe. The School showed an interest in explaining the Universe by natural causes, but Bernard was principally concerned with the reconciliation of Plato's thought with Aristotle's. Bernard's contemporary, Bernard of Clairvaux (1090-1153), was a powerful figure respected throughout Western

Europe for his approach to theological truth through meditation and intuition. His chief work, *De Consideratione,* exerted a restraining force on more enlightened thinking and helped pave the way for scholasticism. Bernard de Cluny (*fl.*1150) was a moralist who expressed disdain for the material world and ascribed reality to ideas and is too close to scholasticism to warrant serious consideration. Bernard de Ventadour (d. 1195?) was a Provençal troubadour who wrote love songs regarded as the finest in his native tongue. Saint Bernard de Menthon (d. 1081?) was the patron saint of mountain climbers who lent his name to two Alpine passes and a famous breed of dog. Bernardo di Pisa (d. 1153) was a disciple of Bernard of Clairvaux and helped launch the Second Crusade. Bernardino da Siena (1380-1444) was a Franciscan theologian who strove to inculcate a deep personal love for Jesus. Bernard of Pavia (d. 1213) was a noted canonist who participated in the *Corpus Juris Canonici* in the interval 1140 to 1500. Bernard of Botone (d. 1263) was another canonist who served as a chaplain to two Popes. Count Bernard VII of Armagnac (d. 1418) died during Paris massacres instigated by Burgundians loyal to the French throne. Bernard of Verdun (13th cent.) was a fanatical supporter of Ptolemaic astronomy. No resume seems especially germane.

Bernardus.

Bernardus Silvestris attracts attention because his writing is relatively free of scholastic influence. Little is known of his life other than that he spent time in Tours, France, where Saxo may have studied. Just as Ben Jonson exhorts readers to learn about Shakespeare by looking at "his Booke," so our knowledge of Bernardus stems mainly from his written works. He wrote about the same time as Bernard of Clairvaux wrote *De Consideratione*, which may have provoked him to satirize regressive thinking. Although he refrained from referring to any of his contemporaries directly, he was a clever and courageous satirist who was unafraid to mirror the foibles of his fellow writers. He was a knowledgeable theologian who was also familiar with contemporary science, and later authors cite his works as examples of a fine literary style.

His poem, *Mathematicus,* addresses astrological prediction and he may have written an allegorical commentary on Martianus Capella's *De nuptiis Philologiae et Mercurii.* His major work, *Cosmographia,* is an excellent fit to the *Hamlet* sub-text. In it, he writes of heliocentric orbits for Mercury and Venus and suggests that the Earth moves. It stands as a landmark of twelfth-century humanism and exemplifies the successful intertwining of science and literature. Like *Hamlet*, it is allegorical and has a moral purpose. Its theme addresses the organization of matter in the Universe, including the creation of humankind. Bernardus believes that the human "is a microcosm of the elements, principles, and forces in the world" and he ends his allegory by recounting the goal of Physis, which is to create the universe of man in imitation of the greater Universe. He believed that humankind was the masterwork of Nature. He associated life with carefully honed universal elements, principles and forces, which anticipates the modern Anthropic Principle that asserts that humans came into being in the Universe of the sort we presently inhabit because the four fundamental constants of nature have values conducive to life.

Cosmographia is unusual in medieval literature because, through allegory, it combines myth and science in service to creation and principles that govern existence. Urania, queen of the stars, looks after the upper reaches of creation and Physis takes care of the lower. Physis is a skilled artisan who works assiduously to fashion humanity according to an inscrutable design, with the result that humans have a divine quality of pure reason and a constitution that conforms to the principles from which it arises. The human capacity for melancholy results from the gravity of the element Earth and Phlegm is the humor that is cold and moist, causing sluggishness and accounting for watery instability. *Hamlet* echoes these properties because Hamlet is melancholic owing to a figurative association to digging in the Earth. He weathers a sea voyage, whereas Ophelia is docile and phlegmatic and dies from a run-in with water.

Plato and other pagan writers inspired *Cosmographia* but there is also a marked Biblical influence. Bernard held high office in the Church and dedicated *Cosmographia* to the Chancellor of Chartres, but his theology is syncretic. He approached natural science in the same way as the School of Chartres, which embodied ancient, medieval and contemporary approaches to the secrets of the Universe. To him, God stands outside the Cosmos while servile angels and pagan deities do all the work. Bernardus would have applauded the famous descent of the deity Jupiter in *Cymbeline* and its return to the heavens, as he would also have approved of the spirit of Old Hamlet receiving its work assignment shortly after admission to supernatural space.

Francisco and Marcellus.

Francisco joins Barnardo in inaugurating the allegorical sub-text and Shakespeare probably uses similar criteria in choosing Francisco's name. A good candidate is the classical Italian poet and philosopher, Francesco Petrarca (Petrarch) whose first name derives from his family's ties to France. He would have impressed the Bard because he is the first Renaissance humanist who rejected medieval Scholasticism. Some say that only Dante's writing surpasses his. Petrarch is a transition figure in literature, and so is Francisco in *Hamlet* because he quickly leaves the stage to Barnardo, a timely exit importing significance to Bernadus' *Cosmographia*. Francisco and Barnardo are soldiers whose alter egos fought against the ignorance of the Dark Ages, as did Palingenius when he exposed the shortcomings of Aristotelianism and lent depth to the distribution of stars. Since Marcellus represents a new vision of the firmament and Horatio hails from Copernican Wittenberg, these two represent the early, un-synthesized, pre-Diggesian components of the New Astronomy. The evidence suggests that Shakespeare introduces and names his characters in the order of their historical significance to the sub-text, so that, when Hamlet enters in 1.2, the forerunners of the New Astronomy congeal about him.

In naming Francisco and Barnardo, Shakespeare may have had candidates in mind other than Silvestris and Petrarca, but this writer could not discover them. At the same time, Shakespeare may not have limited his attention to these two. For example, Petrarch's writings contain the first significant commentaries on the use of eyeglasses after Roger Bacon, and, in 1305, Bernard de Gordon made the first reference in a medical book to spectacles. Piero della Francesca (c.1420-1492) wrote a treatise on geometry and had a mastery of perspective and the properties of light. In the 1600 collection of *Voyages in Search of the*

North-West Passage by Richard Hakluyt (1552-1616), Humphrey Gilbert (1539?-1583), half-brother to Walter Raleigh (1552?-1618), asserts that "Plato, Aristotle, and other famous philosophers" confirm the existence of this sea route to the Orient, as do fourteen contemporary geographers. Among these, Hakluyt lists the names "Franciscus Demongenitus" and "Barnardus," one after the other. Perhaps Shakespeare had in mind Claudius Ptolemy's advances in terrestrial cartography as well.

The Arrow of Time.

Hamlet greets Barnardo, Marcellus and Horatio, and twice asks Horatio what made him leave Wittenberg and come to Elsinore. The repetition suggests that Hamlet is surprised to see him. If Horatio had come to see the funeral of Old Hamlet that had taken place weeks earlier, one wonders how, in such a limited space as Elsinore, the two had not encountered one another sooner. Horatio arrived from Wittenberg on the third night, just in time to see the apparition, and had not had a chance to meet Hamlet until the next day. Horatio would not know of the discovery of the New Star when he was in Wittenberg because that discovery occurred after he left for Elsinore and weary travelers sleep by night. Perhaps, extrasensory perception guides Marcellus and he waylays Horatio before he has a chance to meet his old friend. When Barnardo asks, "What, is Horatio there?" and Horatio answers, "A piece of him," Horatio means that he represents only heliocentricism from Wittenberg and not the new vision of the starry Firmament, which Marcellus represents. It is appropriate for Horatio to arrive escorted by Marcellus also because Horatio comes from the town where Schuler first sighted SN 1572; and withal, in 389 AD, a certain Marcellinus is said to have reported a New Star.

Schuler first saw SN 1572 on the morning of November 6, 1572. At that instant, the New Star phenomenon became part of the cosmic puzzle that the New Philosophy needed to address. Since Hamlet personifies the Diggesian model of the New Astronomy and is the one who will wage the struggle on behalf of the New Philosophy, it seems plausible to associate the time of discovery with the night when Hamlet first sees the Ghost. *Hamlet* says that the New Star was visible for three nights prior to the action in 1.4 and 1.5, so perhaps it was visible for three nights prior to November 6, 1572. This would put the date of first visibility at November 3, 1572, a date that accords with the apparently well-established fact that, on the previous night, a professor at the University of Valencia and his students examined the part of the sky that includes Cassiopeia, and no one saw the New Star. The inference is that Shakespeare – or someone in his immediate circle – saw SN 1572 on November 3 and the Bard incorporated the data in the *Hamlet* allegory.

After Francisco exits, Barnardo and the two new arrivals, Horatio and Marcellus inherit the stage. If characters come and go in the order of their historical significance, we expect that Barnardo will soon drop out of sight because *Zodiacus Vitae* came after *Cosmographia*. What little Barnardo says in 1.2 is uttered in perfect unison with Marcellus – not once but four times – and indeed, this is the last we hear of Barnardo. By 1.4, only Marcellus and Horatio remain to assist Hamlet.

France.

In 1.2, Laertes comes before the king to seek leave to return to France and, within five lines, the king grants his request; but when Hamlet expresses a desire to return to Wittenberg, the court reaches a decision only after a lengthy palaver of 56 lines. Unlike Laertes, Hamlet does not get his way. Among other things, the contrast highlights the difference in destinations, which is confirmed when the next act reveals that Paris, France, is Laertes' precise destination. Paris University was a citadel of scholasticism whose docents regarded Ptolemy's algorithm as the one true system of the World, with some denying even the possibility of other solutions. Shakespeare associates Laertes with Paris in order to anticipate his eventual embrace of Claudius' cause and because the great champion of Aristotelianism, Thomas Aquinas, was a Professor of Theology there. We are not surprised, therefore, when the geocentric court approves travel to Paris but not to Wittenberg.

After the passing of Albertus Magnus and Aquinas, scholasticism underwent a slow decline, which coincided with the emergence of greater objectivity and critical thinking as manifest in the work of Jean Buridan (*c*.1300-*c*.1358) and others. Shakespeare dramatizes the role that Frenchmen played in stemming the tide of scholasticism by choosing the names Francisco and Barnardo for scholars with French connections who initiated enlightened, anti-Thomist thinking. To emphasize that France is as much a part of the political and scientific revolution as any country, Shakespeare names Fortinbras for a Saracen giant in old French literature and assigns him the role of fostering and protecting the New Philosophy.

Harriot, Percy, Raleigh.

After Osric tells Hamlet that the king has laid a wager on his head, there ensues an exchange that most consider ridiculous, puzzling, and confusing. Mowat and Werstine (*Hamlet* 5.2.118-95n) comment, "Often we can only guess at what they might be saying." F1 omits thirty lines of this confusion (5.2.100-25 and 5.2.127-30) because the passages shorten the play and are, supposedly, unessential to the plot. However, the cosmic allegory clarifies their purpose and restores their need.

In 5.2, Osric, Hamlet, and Horatio discuss the weather. Osric's and Hamlet's contrasting comments, "it is very hot," "tis very cold," "it is indifferent cold" and "it is very sultry and hot" refer to *A briefe and true report* on the Virginia colony written in 1588 by the English scientist and mathematician, Thomas Harriot. In it, Harriot compares the climates of Virginia and England, "the excellent temperature of the ayre there at all seasons, much warmer than in England, and never so violently hot" (Hariot 44-5; Quinn 383). Shakespeare lists about two dozen characteristics that supposedly describe attributes of Laertes as enunciated by Osric, but every one also describes a characteristic or a chief accomplishment of Harriot up to the time of Q2. By line numbers, 5.2.100 to 130, these are:

100-1: Osric announces, "... here is newly come to court Laertes." Harriot returned from in 1586 and, after publication of his report, he emerged as a leading intellectual. By 1590, most people recognized him as a foremost English mathematician along with John Dee and Dee's student, Thomas Digges, at which time his work began to be cited. In *c.*1601, therefore, Shakespeare could legitimately write that Harriot was "newly come to court" in the sense that he was recognized by Elizabethan society.

101: Osric continues, "... believe me an absolute gentleman." Notes inscribed on Thomas Harriot's matriculation to Oxford list his age as

17 and the social status of his father as "plebian." Thus, he belonged to the social order of the common man, but his social class changed on graduating BA in 1580 at age 20. It changed further when, in 1595, he received a gift from the ninth Earl of Northumberland, Henry Percy (1564-1632), of a life interest in the income of Percy's holdings in Durham, enabling Harriot to become a lifetime member of the landed gentry. As a result, Harriot was entitled to attach the title "Gentleman" to his name.

101-2: "... full of most excellent differences." "Differences" are characteristics or distinctions that are out of the ordinary (Jenkins 5.2.108n) so that a person "excels in a variety of different accomplishments" (Edwards 5.2.102n). By 1601, select circles recognized Harriot's accomplishments in several fields, including navigation, cartography, ethnography, linguistics, meteorology and his work on atomism, which began with his studies in military science.

102: "... of very soft society and great showing." The phrase "soft society" means "easy sociability," and "great showing" means "excellent appearance" (Edwards 5.2.102n). Harriot's personality was warm and attractive and Percy admired his affability and learning.

103-4: "Indeed to speak freely of him, he is the card or calendar of gentry." He is the map or guide of gentility. A "calendar" is a registry or directory essential to keeping track of events and time. A "card" can mean a map or a stiff piece of paper containing the points of the compass (*OED*). Harriot is both a model gentleman and an expert cartographer and navigator. He kept the maps of Walter Raleigh up to date, especially those of the New World, and drew a map for the Guiana expedition and of Raleigh's Irish holdings. These lines connect the preceding ones on Harriot's gentility, to the next topic, his voyage to the Virginia colony, where Harriot was the official mapmaker and surveyor of the coastline.

104-5: "... for you shall find in him the continent of what part a gentleman would see." On April 9, 1585, a fleet of seven ships, led by Sir Richard Grenville (1542-1591) and his flagship *Tiger,* set sail from Plymouth bound for Roanoke Island. Evidence suggests that Harriot, now a gentleman by virtue of his Oxford education, accompanied Grenville aboard the *Tiger*, and reached the Carolina outer banks in late June.

106-7: Hamlet responds, "Sir, his definement suffers no perdition in you, though I know to divide him inventorially would dozy th'arithmetic of memory." Hamlet agrees with Osric, adding that to list all his qualities would make one dizzy. Such an inventory would have to come from memory for want of a significant number of published works (5.2.108-109 below). The words "divide" and "arithmetic" refer to Harriot's mathematical prowess.

108: "... and yet but yaw neither in respect of his quick sail." To yaw is to swing off course. The fleet led by Grenville on the *Tiger*, having set sail on April 9, 1585, soon encountered a storm that sank the *Tiger*'s pinnace and scattered the fleet. The *Tiger* sailed on alone, reaching the Canaries five days later. It proceeded westward, reaching Dominica in the Lesser Antilles on May 7 and Puerto Rico on May 10. This was "a rapid passage" (Quinn 159) or "a rapid crossing" (Shirley 126). The phrase "quick sail" refers to this rapid crossing and is also a pun on Grenville's "fleet," for at least two meanings of "fleet" were in use at the end of the sixteenth century (*OED*), "a sea force," as in "flete of schyppys," and "swift," as in "fleeter than arrowes" used in *Love's Labour's Lost*.

108-109: "But in the verity of extolment, I take him to be a soul of great article." In truth, "there would be many articles to list in his inventory" (Edwards 5.2.109n), if only he had published them. In fact, the only work Harriot published in his lifetime was *A briefe and true report*. He had raised expectations that he would publish a full account of his research on Virginia, but this large discourse never appeared (Sokol 2). His textbook on navigation *Arcticon,* was never published, either. His reputation as a mathematician results from unpublished papers and a draft of a text (*Artis Analyticæ Praxis ad Aequationes Algebraicus resolvendas*; London 1631), in which he "virtually gave to algebra its modern form" (*DNB*).

109-10: "... and his infusion of such dearth and rareness." That which is "poured into him" by nature is dear and rare (Edwards 5.2.109-110nn). *OED* uses this line to illustrate the meanings of "infusion" and "dearth," but another meaning for "infusion" is the action of infusing some principle or idea into the mind, used as early as *c.* 1450. This fits well with Harriot's dearth of publications since fewer publications imply less impact on people's thinking.

110: "... as, to make true diction of him." Harriot was generations ahead of his time in creating a way to reduce speech to symbols. Unfortunately, he did not leave a treatise on phonetics nor a key to his symbols.

110-1: "... his semblable is his mirror." "The (only) person like him is his own image in the glass" (Jenkins 5.2.118n). Harriot brought several scientific instruments with him to Virginia, including "a perspective glasse whereby was shewed manie strange sightes" (Quinn 375). Harriot therefore possessed and demonstrated what is generally regarded as the forerunner of the telescope (Quinn 375n4), but it was not until after *Hamlet* was written that he studied celestial objects telescopically.

111: "... and who else would trace him." Here "trace him" means "follow him closely" (Edwards 5.2.111n). After the storm scattered the fleet, the *Tiger*, presumably with Harriot on board, arrived at the appointed rendezvous eight days ahead of the next fastest vessel. No ship of the fleet followed the flagship closely. The present interpretation supports the conjecture that Harriot accompanied Grenville aboard the *Tiger* (Shirley 125-6).

111: "... his umbrage." At sea, Harriot observed a partial eclipse of the Sun and was partly in the Moon's "umbrage" or shadow.

112: "... nothing more." Shirley (125) writes, "whether he attempted to use these observations [of the eclipse] to calculate longitude accurately is doubtful." Harriot passed up this opportunity to contribute to the solution of a difficult problem.

113: Osric is pleased with Hamlet's flattery. "Your lordship speaks most infallibly of him," he says. Here, as in 5.2.81 and elsewhere, Osric and Horatio address Hamlet as "lord" but Horatio and Osric are addressed as "sir" (see 5.2.117-8 below).

114-5: Hamlet seeks relevancy and asks, "The concernancy, sir? Why do we wrap the gentleman in our more rawer breath?" In other words (Jenkins 5.2.122n, 122-3n), "How does this concern us? Why do we clothe him in words of ours which can only fall short of his refinement?"

116: Osric is baffled. "Sir?" he asks.

117: Horatio tries to be helpful and asks Osric, "Is't not possible to understand in another tongue?" The 1584 Amadas-Barlow expedition to Virginia brought back to England two American Indians, from whom

Harriot learned Algonquian and then developed a special alphabet by which to record their language. The line may also refer to the facts that, with the help of Richard Hakluyt, Harriot's *A briefe and true report* was published in three languages besides English, and to Harriot's fondness for Latin at a time when English was emerging as the language of scientific literature.

117-8: Horatio mimics Hamlet's balderdash, "You will to't sir, really." Horatio always addresses Hamlet as "lord" (Jenkins 5.2.125-6n) and here addresses Osric as "sir." As it stands, the remark makes little sense and since Horatio does not explain to what "it" refers, we are free to improvise. At a stretch, perhaps he means, "You [are] Walter Sir, Raleigh," where "will to't" and "really," suitably slurred, sound like "Walter Raleigh" (but cf. Plumptre's solution, Chapter 5.) "You" then suggests that Horatio identifies Osric as Raleigh (see Table 2). The possibility is credible given that Raleigh championed Harriot just as Osric champions Laertes. Moreover, at the outset of their meeting, Hamlet identifies Osric to Horatio as one who "hath much land and fertile," which fits since Raleigh owned much fertile land (103-4 above).

119: Hamlet tries again. "What imports the nomination of this gentleman?" he asks. He wants to know "the purpose of naming this gentleman" (Edwards 5.2.119n), i.e., of naming Raleigh. The question answers itself thanks to a pun on "imports." According to the *OED*, "import" means a commodity brought in from abroad and was in use only by 1690; but the associated verb was in use much earlier, in 1508 and 1548. Perhaps the remark refers to Raleigh's interest in importing natural resources from England's colonies. For example, in *A brief and true report,* Harriot makes "declaration of such commodities there alreadie found or to be raised, which ... by way of trafficke and exchaunge with our owne nation of England, will enrich your selves the providers: those that shal deal with you; the enterprisers in general ..." (Quinn 324; Shirley 146). To make his point, Harriot returned to Raleigh a collection of specimens from the West Indies.

120: Osric is still confused. He asks, "Of Laertes?" He does not realize that Hamlet has shifted the conversation briefly to Raleigh (i.e., to Osric himself). By asking this question, Osric shifts attention back to Laertes, i.e., in the present reading, back to Harriot.

121: Horatio responds accordingly, "His purse is empty already, all's golden words are spent." "His purse" refers to H[is]enry P[urse]ercy's coffers, from which gifts of money were made to Harriot. In the early 1590s when Raleigh was in disfavor with Queen Elizabeth, Harriot sought other patronage; and it was natural for him to turn to Raleigh's friend, Percy. Percy's largesse commenced in 1593 with a sizable annual grant of £80. Thus, the names of the two persons who significantly influenced Harriot's life, the close friends Raleigh and Percy, appear in the order in which they patronized him. Although "[H]is [P]urse" refers to Henry Percy, the second half of the sentence suggests that his purse might store "golden words" rather than golden coins, so that in this case "H[is]" refers to H[arriot] as well who has a purse full of words at his disposal. As coins are held in a purse before they are spent, so "his purse" could also connote H(is)arriot's P(urse)ercy. Despite Percy's generosity, by 1601, Harriot had produced only one work and showed no sign of producing more. The line contrasts the cessation of Harriot's output after 1588 with his receipt of a lavish pension from Henry that resulted in no published works. It seemed as if Harriot had depleted his store of words and had none left to commit to paper. Shakespeare may have intended the Osric dialogue to be highly opaque because, in about 1601, Harriot, Raleigh, and Percy, were alive.

122: Hamlet says, "Of him sir." The words "of him" have occurred three times above (103-4, 110, 113), and the referent in all cases is Laertes, i.e., H(im)arriot. Horatio's remarks refer to Percy, and Hamlet says that they refer also to Harriot.

123: Says Osric, "I know you are not ignorant ..." Osric (a.k.a. Raleigh) is referring to the prologue to Harriot's *A briefe and true report*, where thrice in the space of six paragraphs Harriot calls his compatriots "ignorant" (Quinn 321-3). Perhaps Shakespeare intends that the word "you" be emphasized, in which case we understand the remark to mean that Osric (Raleigh) knows that Hamlet (Thomas Digges) is not ignorant; the presumption being that Laertes (Harriot) is not ignorant, either. Shakespeare's intent is to show that the quoted remark is tantamount to Raleigh (Osric) telling Thomas Digges (Hamlet) that he, Digges, is not ignorant!

124-5: Hamlet interrupts, "I would you did sir, yet in faith if you did, it would not much approve me. Well sir?" Hamlet has no need for fatuous compliments from Osric any more than Thomas Digges needs them from Walter Raleigh. What does Osric have to say about that?

126: Osric tries to recover by explaining, "You are not ignorant of what excellence Laertes is." Osric supposes that Hamlet (a.k.a. Thomas Digges) knows all about the excellence of Laertes (a.k.a. Harriot), but Osric has dug his hole a little deeper.

127-8: Hamlet cannot admit to such a comparison, "I dare not confess that, lest I should compare with him in excellence, but to know a man well were to know himself." For Hamlet to admit knowing of Laertes' excellence would be like claiming such excellence for himself, since to *know* such excellence one would need to be able to *perform* such excellence; beyond that, the sentence "is not meant to have much meaning" (Edwards 5.2.127-8n). On the contrary, the comparison between Laertes and Hamlet is one between Harriot and Thomas Digges and, as far as Hamlet is concerned, it is an unequal comparison.

129-30: Osric narrows the comparison to weaponry, "I mean sir for his weapon; but in the imputation laid on him by them, in his meed he's unfellowed." Undeterred, Osric continues to press the cause of Laertes, but Hamlet has had enough. The unexpurgated text resumes with Hamlet's pointed question, "What's his weapon?"

Shakespeare gives Harriot special attention because both he and Digges opened up new frontiers. Digges saw the stars and described the new World view, and Harriot saw Roanoke and described the New World. Harriot unearthed a new continent, and Digges unveiled a new cosmos. The new vistas are complimentary inasmuch as they helped transform understanding of Earth and Sky. In 1533, both geographic and cosmographic frontiers were at the forefront of English consciousness as depicted, for example, by the items on the upper and lower shelves in Hans Holbein's painting *The Ambassadors,* and by the fact that the gentlemen of the Inns of Court set great store by both topics.

Portraiture.

Cosmic sciences are chief characteristics of the *Hamlet* allegory, yet the items attributed to Harriot in Q2 barely mention them. Harriot was lax in not publishing his work despite his obvious talents and, at the turn of the century, he was, therefore, in no position to demean workers as "ignorant." After Q2 and after 1603, Harriot's science flourished as James VI began to affect the lives of Harriot's benefactors, Raleigh and Percy, and he then became seriously interested in applying his optical theories to the study of the heavens. In about 1609, well after Q2 had appeared, Harriot developed a refracting telescope at about the same time as Galileo and, like Galileo, trained it on the Sun, Moon, and planets. Harriot discovered sunspots independently and deduced the Sun's period of rotation. He collected data on the Comet of 1607 (Halley's) that Bessel later used to write his first paper. In 1609, he made the world's first telescopic drawing of the Moon. He observed the Comet of 1618, and he accomplished all this in the space of less than 20 years after the end of Elizabeth's reign. At the time of his death, two years before publication of F1, Harriot's scientific reputation was high and *Hamlet* needed revision.

In Q2, Shakespeare was less than complimentary toward Harriot. The depreciatory term "rabble" characterizes those who support Laertes a.k.a. Harriot and probably refers to Raleigh, Percy and those many burghers of England who thought that Harriot was their leading scientist. F1 changes all this. Deletion of this deleterious material means, in effect, that Shakespeare's editors or collaborators had second thoughts. "Shakespeare" apologizes to Harriot by having Hamlet admit that he was rude to Laertes. "I am very sorry ... that to Laertes I forgot myself," he says, explaining that he now sees the "portraiture" of Harriot's cause in the "image" of his own, which is, reasonably, to advance knowledge while avoiding charges of sedition and atheism. It is likely that both Harriot and the Bard feared harassment and reprisals. Harriot did time in the Gatehouse as an indirect result of the Gunpowder plot of November 5, 1605, albeit only because untoward circumstances tarnished his mentor, Henry Percy. These events occurred after publication of Q2, following which Shakespeare's collaborators or editors recognized in Harriot the perilous state of all practitioners of the New Philosophy.

The apology appears in F1 at a point in the text almost immediately before Hamlet enters into the disparaging dialogue of Q2, which already contains Hamlet's earlier attempt to seek forgiveness. Hamlet tries to explain it was not he who wronged Laertes but his "madness." The allegorical model readily explains this first attempt, but the additional apology has a different tenor. In so emending the play, the editors of

F1 also remove passages offensive to the memory of Raleigh, knowing that, in 1614, Raleigh had authored a notable *History of the World*.

Cornelius and Voltemand.

Reference to the New Philosophy occurs also in the names that Shakespeare assigns to Cornelius and Voltemand. Sohmer (*Mystery* 222, 242n8) points out that Marcellus is the name of an early convert to Christianity, that Voltemand means, literally, a "turned" or a "changed man" and that Cornelius is the name of a Roman centurion to whom an angel of God appeared and ordered him to seek out Simon Peter. Peter preached to a crowd and the gentiles were astonished to find that the gift of the Holy Ghost poured on them as well (*Acts* 10). If Shakespeare were ever to run into trouble with censors or bring England and her poets into disrepute, canonical religious themes would serve as a fallback position by which he could plausibly deny heresy. Perhaps the Bard was in no mood to become living proof of the adage that history repeats itself. No one has yet located any book or manuscript belonging to Shakespeare, suggesting that he went to great lengths to protect his worksite – and himself. He was careful to protect the identities of his sources of astronomical information as well, as is again evident from the puzzles unraveled in the next Chapters.

CHAPTER 10: THE NEW ASTRONOMY

The most fateful human invention was the lens. The lens destroyed
the world into which it was put.

Wendell Johnson

Authors have wondered whether Digges studied the stars
telescopically, but have neglected to ask whether he looked at the Sun,
Moon, and planets. Stars are common celestial objects distinguished by
brightness and color, but planets attract attention in other ways and are
just as intriguing. The proposition deserves testing.

Omens.

In 1.1 of Q2, Horatio speaks of dreaded events:

> In the most high and palmy state of Rome,
> A little ere the mighty Julius fell,
> The graves stood tenantless and the sheeted dead
> Did squeak and gibber in the Roman streets.

These echo the words of Caesar's Calpurnia that "graves have yawned
and given up their dead" and "ghosts did shriek and squeal about the
streets." Horatio continues (1.1.117-20):

> As stars with trains of fire, and dews of blood,
> Disasters in the Sun; and the moist star,
> Upon whose influence Neptune's empire stands,
> Was sick almost to doomsday with eclipse.

These are "harbingers ... and prologue to the omen coming on" (1.1.122-
3). Interpretation of these passages is difficult if only because 1.1.117-8
are incomplete. However, "Dews of blood" does immediately precede
and belong with "disasters in the Sun," suggesting a relationship between
blood and the royal Sun.

The impending calamities involve the Sun and a lunar eclipse
and the ominous tone forewarns of the deaths of Hamlet and Ophelia.

Hamlet figures prominently in the next scene (1.2) and Ophelia in the one thereafter and both omens are associated with celestial excess. Hamlet's being too much in the Sun is a pun on his being "too much" the "son" of Old Hamlet; but an equally apt interpretation of this solar excess concerns the integrity of the royal body. In "disasters in the Sun," the word "disasters" is a plural noun, referring to multiple events of some kind that involve the Sun. "Disasters in the Sun" are likely not analogous solar eclipses because the solar "disasters" are plural whereas the lunar "eclipse" is singular.

Plasma.

Perhaps the phenomenon of sunspots serves as Hamlet's omen. Sunspots can exceed tens of thousands of kilometers in size, large enough to encompass several Earths. The larger ones are readily visible to the naked eye when atmospheric conditions reduce the sun's glare. Sunspots form because the Sun's outer layers are in a state of convective turbulence, which generates magnetic fields. Sometimes the fields become so strong that they burst out from the deeper layers and affect the appearance of the visible surface. Magnetic fields inhibit the motion of plasma and thus diminish convection, so the surface gases cool because the hot plasma from below does not replace the cooler surface plasma at the same rate as occurs elsewhere on the surface of the Sun. The resulting areas are cooler than the surrounding photosphere. Small temperature differences give rise to large changes in radiant flux so that, to the naked eye, sunspots appear dark and black, like holes in the Sun's surface. In fact, William Herschel (1738-1822) believed that sunspots actually were holes in the Sun through which observers could see the solar interior, cold and dark as Erebus. At that time, no one realized the physical impossibility of insulating the solar interior from its surface heat.

Holes.

Sunspots were known to Charlemagne (742-814) and possibly, to Virgil, and Ptolemy was said to have seen them too. In about 1612, Galileo determined that sunspots are associated with the solar surface and are not some sort of interplanetary phenomenon. He and Harriot observed their relative angular speeds across the face of the Sun and saw that spots always move less rapidly near the rim than near the meridian. This difference results from foreshortening, as spots near the rim of the rotating Sun have less motion across the line of sight than those near the meridian where lateral motion is greatest. By contrast, a passing foreground object would move uniformly across the apparent disk of the Sun.

By speaking of disasters *in the Sun*, Shakespeare probably means that they are actually *part of the Sun* itself and not some foreground phenomenon. Even if the latter were true, he probably would not believe that a transit of some dark body, like the planet Mercury, portends disasters on Earth since there is no analogous superstition or lore. The significance of the preposition "in" in "disasters in the Sun" is apparent also when Hamlet says to Horatio that many things exist in heaven and *in earth*, not *on* Earth. Since sunspots look like holes *in* the skin of the Sun and since the Sun represents kingship both in the sky and on Earth, one might expect that holes pertain to all claimants to the throne. In fact, Claudius, Hamlet, and Laertes all suffer puncture wounds.

Shakespeare is unlikely to engage in guesswork and must have had excellent empirical evidence if he meant us to take *in the Sun* literally. The word "in" carries weight disproportionate to its length, but we are not dismayed because Shakespeare often makes weighty pronouncements with little fanfare. The present hypothesis suggests that the lunar eclipse and solar sunspot disasters together foretell the deaths of Ophelia and Hamlet, the former by drowning, the latter by wounding.

Cytherea.

From the outset, Ophelia's father and brother oppose her romantic involvement with Hamlet. In 1.3, just before he departs for France, Laertes inundates Ophelia with advice. He speaks of Hamlet's amorous interests and advises her not to take matters of the heart too seriously in light of Hamlet's youthfulness. In keeping with mythology, Shakespeare relates love and chastity to their respective deities, love (1.3.5-28) to Venus, and chastity (1.3.29-44) to the Moon. Laertes begins by addressing the subject of love in general (1.3.11-14):

> ... nature crescent does not grow alone
> In thews and bulk, but as this temple waxes
> The inward service of the mind and soul
> Grows wide withal.

In other words, growing up is not just a matter of physical size because, as the body grows, so the inner life of mind and soul must develop too.

The word "crescent" pertained originally to the waxing Moon regardless of phase, but, by 1578, it had come to mean the illuminated shape when the Moon is less than half full (*OED*). In the passage above, most editors avoid connecting the word "crescent" to the Moon, no doubt because they know that the issue is love, which is the province of Venus, whereas, as noted, chastity is the province of the Moon. One editor throws caution to the wind (Martz 1.3.11n) perhaps because, before now, it had been inconceivable that Shakespeare would make such a reference to any celestial body but the Moon. The implicit assumption is that, in the sixteenth century, the Moon was the only Ancient Planet known to have crescent phases, but Shakespeare makes it abundantly clear that the immediate context is love, not chastity, and, in so doing, he intentionally associates the process of waxing with Venus, not the Moon.

Shakespeare's description of a crescent shape agrees with the description of a *change of shape* in which "love" (i.e., Venus) "waxes" and "grows wide." The waxing process involves phases that change from New, via crescent, to First Quarter, and then to gibbous and Full, this order being reversed in the waning phase. Shakespeare refers to the complete waxing phase from crescent to gibbous as a "widening." He did not use the term "gibbous," which refers to the phase between quarter and full, since this technical term entered the English language only in 1690 (*OED*).

Since the phases of Venus correlate with its position with respect to the Sun, Venus shines by reflecting sunlight just like the Moon. A further consequence is that Venus shows a full range of phases as it "grows wide" only because it circles the Sun, in the course of which observers on Earth see it from different angles. Galileo reached the same conclusion in 1610, after he observed all phases with his spyglass. Venus would not manifest a full range of phases if it orbited the Earth and, if it were to lie always between the Earth and the Sun, it would never show gibbous phases, and conversely, if it were always farther away than the Sun, it would never show crescent phases.

Shakespeare would not have asserted positively that Venus shows the full gamut of phases without empirical evidence. If not telescopically, could the naked eye have provided it? The appearance of Venus is best discerned when its angular size is largest, i.e., when it is close to Inferior Conjunction, yet sufficiently far from the Sun in angular distance to render it visible in the twilight. The optimal angular size falls in a range of about ½ to 1 arc minute, which is the stated range of the minimum resolvable angle for ordinary visual acuity corresponding to 20/10 or 20/20 vision. Resolution could exceed this when the eye seeks to locate an element relative to another of the same target, in which case the so-called hyperacuity of the eye can have a minimum discriminable angle as small as 2 to 10 arc seconds. Such detections occur through complex neural processes that would need quantification if visual data were to have scientific value, but, in antiquity, there were no studies to assess the capabilities of human eyes and no controlled experimentation to quantify effects of astigmatism, near- and far-sightedness, weather conditions, seeing, and glare.

Under optimum conditions, the naked eye might barely resolve the image of Venus, and early Babylonian accounts do report that Ishtar (Venus) has "horns," a shape also seen by keen-sighted contemporary observers. Perhaps, these observations entered lore and manifested themselves in ancient culture. Conceivably, the combined crescent and starlike images on the flags of Muslim countries could represent Venus near Superior and Inferior Conjunction. Moreover, it is more likely that the two images are different representations of the same object than that they are pictures of the crescent Moon and some otherwise unspecified star.

Even if the characteristic mark of Muslim nations is Cytherean in origin, there is no evidence that such information became part of lore in Europe. Galileo used an anagram, *Cynthiae figuras aemulatur amorum* ("the mother of love [Venus] emulates figures [phases] of [the Moon] Cynthia"), to establish priority for his discovery of the phases, but would not have done so if the fact were known. If Shakespeare were to have based his text on lore, he would surely have left textual clues to that effect, but none is evident. Neither is there any historical evidence for a *progressive change* in phase that Shakespeare describes ("grows wide withal"), suggesting that his positive assertion is most likely based on the quality of information that a telescope supplies. It seems reasonable to conclude, therefore, that spotty and generally unverified evidence on Venus from antiquity would not explain the textual clarity of the astronomical descriptions with which Laertes harangues his sister.

The range of phases for Venus is consistent with the Copernican hypothesis but does not prove it because a heliocentric orbit for Venus also supports the Egyptian-Capellan-Tychonic solution. To make the case for the New Astronomy, Shakespeare needs more grounds than this.

Cynthia.

Having declaimed on love in general, Laertes turns next to love and its possible consequences. He starts and ends the passage 1.3.33-44 by urging Ophelia to remain a virgin, "Fear it Ophelia, fear it my dear sister" followed by "best safety lies in fear." He warns of dishonor if she were to entertain Hamlet's advances seriously or open her chaste treasure to the swain's unmastered importunity.

Laertes tries to instill fear in her with the help of three metaphors whose common theme is physical impact. The first is a purely military metaphor urging Ophelia to stay out of the "shot and danger of desire." Next, Laertes cites two adages. The first says that a maiden would go far enough if she were merely to reveal her beauty to the chaste Moon, and the second describes the threat to Ophelia's virtue, "Virtue itself scapes not calumnious strokes." This is a variant of the sayings, "Envy shoots at the fairest mark" (Hibbard *Hamlet* 1.3.38n) "Calumny will sear Virtue itself" from *Winter's Tale* and "calumny / the whitest virtue strikes" from *Measure for Measure* (Jenkins 3.1.137-8n).

Last, Shakespeare launches into a botanical conceit, likening the possibility that Ophelia might have a serious relationship with Hamlet to the cankers that afflict young plants:

> The canker galls the infants of the spring
> Too oft before their buttons be disclosed,
> And in the morn and liquid dew of youth
> Contagious blastments are most imminent.

This is a variant on the saying, "The canker soonest eats the fairest rose" (Tilley C56). Theophrastus (372-285 BC) knew of plant canker and that it manifested itself as roundish spots or lesions. These result from injury to the leaves, stems, thorns, and fruits of plants, allowing fungi and bacteria to assail the tissue. A diseased fruit has sunken craters and a raised margin surrounding the cavity where the plant resists the infection, causing it to resemble lunar craters. The contagion is especially harmful to young plants.

"Gall," as a noun, has been in use since 1398 to mean a type of excrescence produced on trees. The verb has been used since 1548 to signify injury by contact and, more particularly, to "harass or annoy in warfare," especially by "arrows or shot" (*OED*). In early use, the word "gall" meant also a "breach" as in "gaules made by the artillerie." Thus, Shakespeare enters into botanical-cum-military wordplay by conflating breaches in two sorts of defenses.

"Blastment" is the first such instance of its use. Edwards (1.3.42n) interprets it botanically as a "blighting," but this meaning is too narrow. "Blastment" seems as descriptive of explosions as of lesions. The context is the chaste Moon, suggesting that blastments have caused her to lose her virtuous perfection and that blemishes scar her surface. Shakespeare is describing craters on the moon, and Laertes advises Ophelia repeatedly to fear them. When Laertes says, "blastments are most imminent," he warns of consequences if they were to occur on Earth. Primarily, the word "imminent" connotes something that is "impending threateningly" and "hanging over one's head." It also means (*OED*) "close at hand in its incidence," and indeed lunar features are "most imminent" since the Moon is the closest celestial body to Earth. The more literal sense (*OED*) of "overhanging" or "hanging over one's head" is appropriate as well because that is where we see the Moon a lot of the time. Even the literal meanings of "projecting or leaning forward" might apply because the Moon peers over the horizon as she rises and looks back as she sets.

The terms "shots," "strokes" and "blastments" describe events akin to powerful explosions on Earth that Shakespeare would know about because he acquired his military knowledge largely from *Stratioticos*. He might also have known of a *Treatise of Great Artillerie and Pyrotechnics* mentioned at the beginning of *Stratioticos* as one of many works by Thomas Digges whose publication never materialized. In the sixteenth century, Shakespeare's astronomer knew the size and distance of the Moon and would know that crater diameters could exceed the distance from London to Stratford-on-Avon.

Plutarch described the Moon as Earth-like and having mountains that cast shadows. Is the Bard simply parroting this description? Although a casual observer on Earth can see that the lunar terminator has a markedly ragged appearance and might ascribe this, reasonably, to shadows cast by a rough terrain, he would not know that the cause of some of that roughness could be inimical to life if it were to occur on Earth. The descriptions leave little doubt that the Bard needed telescopic observations to construct his lunar and Cytherean conceits.

Pox.

In 1988, Charney (120-30) devoted a full chapter of his book to what he termed the largely neglected subject of skin disease in *Hamlet*, stating that this imagery is the most distinctive of the play. The fact that pockmarks on the Moon are visible only with optical aid and that sunspots themselves are like epidermal lesions, supports Charney's interpretation. Cankers with raised rims, like those on diseased fruit, are characteristic of syphilis. After sailors first returned from the New World in 1493, syphilis ravaged Europe although, by the late 1550s, it had "settled into a life of a subtle three-stage microbe" with an incidence so ubiquitous that it even afflicted the ostensibly celibate. The only defense was chastity, making Ophelia's an issue for Laertes.

The Bard devotes over 180 lines in the Canon to the subject, with additional commentary in the Sonnets. Ross posits that the Bard might have suffered from the disease and from the effects of its treatment. He suggests that William Shakspere insulted Anne, nee Hathaway, in his infamous bequeathal to her of his second-best bed because he had experienced other beds more to his liking. Perhaps Shakspere caught the disease there, but, of course, this is not to say that Shakespeare did.

Shakespeare uses the word "pox" in exclamations of irritation or impatience, and "plague" occurs in *Henry IV part 1* and *Romeo and Juliet* ("A plague on both the Houses"). In *Hamlet,* he uses "pox" to express annoyance with the Player Lucianus and, significantly, he mentions "faces" in the same sentence ("Pox, leave thy damnable faces and begin"). This expression of irritation relates to the contagious lunar blastments because craters cover the visible face of the Moon much as the ulcers of smallpox blemish the human face. In 3.1, Hamlet says that Polonius should confine himself to his home, and he goes on immediately to mention "this plague." By the late fourteenth century, any disease that corrupts or infects was termed contagious. Earlier, by the turn of the fourteenth century, to "corrupt" could mean to render morally unsound (*OED*). The image of moral corruption and the pockmarks that beset life combine with astronomy and mythology to describe Ophelia's hypothetically unchaste state. When Ophelia tries to return letters to Hamlet that were supposedly written by him, Hamlet denies having given any to her, prompting him to question her fairness and honesty. In the sixteenth century, "honest" and "fair" meant "without blemish." Mindless delirium is a symptom of syphilis, which would account for Ophelia's condition just prior to her death.

By contrast, Hamlet is "indifferent honest" or "moderately virtuous" (Edwards 3.1.120n) in the sense of "equal ... identical, the same" (*OED*). Hamlet admits that he is in the same boat as Ophelia and destined for oblivion as well. Like her, he is blemished, the implication being that the Sun is blemished like the Moon, as indeed it is, by sunspots. The two defects, Ophelia's dishonesty and Hamlet's indifferent honesty, are paralleled by the existence of different kinds of blemish in their celestial counterparts. Hamlet's meaning is ambiguous for he could mean that he is "disinterested," or "unconcerned" (*OED*), which in the context of the sub-text means that his honesty is beside the point because his primary duty is to the mighty lawgiver of the Universe. Hamlet tells Ophelia that she shall "not escape calumny" even if she had been as chaste as ice and pure as snow. There is a nuance of a difference in the honesty of each, suggesting that Shakespeare regards lunar craters as more hideous than sunspots since the former are permanent and the latter ephemeral. The upshot is that Ophelia is imperfect, perhaps in sundry ways, and is a suitable human counterpart to her maculate celestial luminary. Hamlet, on the other hand, pursues a murderous vengeance that, some might say, plumbs the depths of human depravity but which, in the sub-text, is immune to moral judgment.

Jupiter and Mars.

If Shakespeare goes to these lengths to describe the appearance of Sun, Moon and Venus, he is inviting examination for descriptions of related phenomena. For the sake of completeness, we should oblige.

In his mother's chambers, Prince Hamlet describes his father's image as it appears in a miniature portrait:

> See what a grace was seated on this brow;
> Hyperion's curls, the front of Jove himself,
> An eye like Mars, to threaten and command.

Here the word "eye" is the object of the imperative "see," as are "Hyperion's curls" and "the front of Jove." It follows that the "eye" refers to Old Hamlet. The phrase "an eye like Mars" is sandwiched between "Jove" (i.e., Jupiter) and his capacity "to threaten and command," this latter attribute alluding to the god of laws who maintains order among the residents of Mount Olympus. An uncritical reading might lead us to believe that "an eye like Mars" means that Old Hamlet has an eye that is like the eye of Mars, i.e., an eye like "Mars's [eye]." However, mythology pays no attention to the God of War's eye, although undoubtedly he has one. There is no apostrophe in the text, and Bullough (34) confirms that it is Old Hamlet's eye that is "like Mars."

Appurtenances of war, like sword and shield, characterize the god Mars, and the planet of that name is reddish and aptly named for the color of carnage. Thus, when Shakespeare writes that Jupiter has an eye that resembles Mars, he implies that the planet Jupiter has a feature like a red eye. In 1664-5, Robert Hooke (1635-1703) and Jean Dominique Cassini (1625-1712) reported a Great Red Spot (GRS) on Jupiter, so, at the turn of the turn of the sixteenth century, it was entirely new to say that Jupiter has an eye that is like Mars. The apparent disc of Mars varies in size from 3.5 to 26 arc seconds, and when the GRS is on Jupiter's meridian, its apparent size is in the range of about 3 by 6 arc seconds to 5 by 10 arc seconds. Both ranges are below the ordinary visual acuity of the human eye, requiring magnification in order to distinguish them. Just as we recognize a person because we can resolve facial features like a nose, mouth or eye, so Shakespeare's astronomer can resolve the Jovian "eye" and the planet Mars and see that they are red.

Moreover, Shakespeare fails to report phases for either planet, implying that, if these planets reflect light like Venus, then we see them always face on. This observation is consistent with Copernican theory because both are Superior Planets and, thus, never have crescent phases. The Jovian feature is aptly termed an "eye" because, as Jupiter rotates, the GRS disappears for more than 5 hours out of every 10, as if an otherwise undistinguished Jovian surface has a blinking "eye."

The GRS lies in Jupiter's Southern Hemisphere. It is a tempest of immense size, large enough to hold several Earths, and unusual in that it is an anti-cyclone, the opposite of the common, low-pressure typhoons of Earth. High-pressure systems are associated with fair weather, allowing outsiders to see deep into the atmosphere. In the GRS, exotic molecules made from hydrogen, carbon, nitrogen, oxygen, sulfur, and phosphorus, lend color to the spot, which varies from a barely perceptible discoloration through orange-red to red. The ferocity of the storm is abetted by the planet's relatively large size and rapid rotation and, as far as we know, the storm has been blowing ceaselessly ever since the Hooke-Cassini announcements of 1664-5. The present interpretation increases the age of the GRS to over four centuries.

"Hyperion's curls" may signify that the hair of Old Hamlet is comprised of ringlets of a sunny color. "Hyperion" is the progenitor of, and sometimes synonymous with, Helios the Sun. "Curl" means a lock of hair or a ringlet and is "applied indifferently to a flat spiral ... a (helix) or anything intermediate to or approaching these forms" (*OED*). The verb "to curl," means to bend round, wind, or twist into ringlets. It does not necessarily connote a helical form or even a shape that closes in on itself, as in *Henry IV part 2* where Shakespeare describes the action of wind on the surge's billows, i.e., on ocean waves that curl their monstrous heads (*OED*). "Hyperion's curls" may therefore refer to some phenomenon of the Sun that has a shape like that of the top of a breaking wave. The words do not refer to a crescent shape that often mistakenly describes the portion of the Sun visible during a partial eclipse. That term is imprecise because the concave side of the solar sliver seen near totality is not elliptical, but circular, and, therefore, is not the same shape as the terminator of a crescent Moon.

On balance, could Old Hamlet have been a curly-haired redhead with eye color to match? As Old Hamlet represents Leonard Digges, so it is worth wondering whether we now know something of the appearance of the father of astronomical telescopy. Perhaps he is the subject of the mysterious portrait of 1588 by Nicholas Hilliard (1537-1619) that shows a comparatively young man, in his middle to late thirties perhaps, receiving the muse of the new organon from the god, Apollo. Leonard was about 37 in 1558, at which time Elizabeth became queen and he redeemed his lands and was free to resume research after the debacle of Wyatt's Rebellion.

Jupiter's Moons.

In 1610, Galileo discovered four bright moons circling Jupiter and reported them in The Starry Messenger (*Sidereus Nuncius*). These moons are practically visible to the naked eye and, in theory, could have been an easy target for the perspective glass. Shakespeare did not report their existence in *Hamlet*, however, suggesting that Digges did not know of them.

Textual evidence supports the proposition that Shakespeare, or his editors or collaborators, reported Galileo's discovery in *Cymbeline* (Usher "Jupiter"). Posthumus is sleeping fitfully the night before his scheduled execution and has a vivid dream in which ghosts of his immediate family summon the Olympian lawgiver, Jupiter, to save him. Three coincidences are noteworthy. First, the number of ghosts circling the stage and appealing to the god, Jupiter, happens to equal the number of bright moons that Galileo discovered orbiting the planet, Jupiter. Second, independent estimates put the date of *Cymbeline* as 1610, the same year as Galileo announced his discovery in "The Starry Messenger." Third, the god Jupiter is effectively a "starry messenger" when he descends from the heaven.

In The Starry Messenger, Galileo also described the lunar surface and the existence of stars below the threshold of naked eye visibility, but *Cymbeline* does not mention these discoveries for the simple reason that Shakespeare had already reported the results a decade earlier, in *Hamlet*.

10,000 Stars.

In 3.3, Rosencrantz describes how "huge spokes" reach outward and support "ten thousand lesser things." One interpretation is that observers with keen eyesight under a dark sky can see stars to apparent magnitude 6.5 and that the sum of all such stars happens to approximate that number. Neither Hipparchus nor Ptolemy estimated the incidence of stars visible to the sensitivity limit of the naked eye, nor, as far as we know, did anyone in the heyday of Muslim science. Tycho Brahe did not attempt the task since he was more interested in measuring the positions of stars; he tried to get positions for 1,000 stars so that his work might appear equal to ancient efforts, but was able to list only 777 with the accuracy he desired. Other sixteenth-century efforts included celestial cartography, which also required knowledge of star positions, but none came close to the relatively large count of 10,000.

The Bard's 10,000 might mean, simply, "a large number," but we suspect that Shakespeare knew that counting stars has little value unless accompanied by data analysis and theory building. Digges and Galileo both reported seeing stars stretching out as far as their glasses could descry, which was, in itself, a great discovery and, in Digges case, may have been sufficient to warrant a leap to a stellar "infinity," but these observations alone do not account for the number 10,000. If Digges determined the number of visible stars, he did not mention the result in *A Perfit Description* perhaps because, in 1576, he had not started or completed counting. However, by the time *Hamlet* came to the fore, he may have finished doing so and Shakespeare would have felt free to report the information.

To get the number of naked-eye stars in the sky, Thomas Digges would need to assume that the sky unseen from England is no different in gross properties from that visible from England. He could then count stars in small, randomly selected patches of the sky that he could see and whose solid angles he could measure, and in principle, from these data he could estimate the total number across the entire sky. The Bard's 10,000 would then be a bona fide estimate and the first quantified statement of the stellar revolution.

Tears.

If Digges did count the visible stars, his result would rest in part on his ability to resolve faint stars and stars of the Milky Way. In 2.2, Shakespeare reports that this nebulous band results from overlapping star images as he lets Player I recite a passage about the savagery of Pyrrhus. It is so moving as to engender sympathy in the gods "unless things mortal move them not at all." If gods were capable of caring, such sympathy would evoke their passion and "make milch the burning eyes of heaven," whereupon Polonius comments that Player I has tears in his own eyes, which shows how heart-rending the description is. In describing how telescopy sharpens the appearance of nebulosity into stars, Shakespeare implies that the stars themselves are not resolved because there is no parallel description of stars being intrinsically nebulous, i.e., resolved.

Polonius' observation suggests that, if the gods were to have seen this brutality, they, too, would have become bleary-eyed. Tears blur vision, effectively causing a loss of optical resolution. Discrete, starry images spread out and overlap, giving a milky appearance. Conversely, when the tears dry up, the opposite occurs. "Milch" refers, reasonably, to the Milky Way (Edwards 2.2.475n) which appears milky because human eyesight cannot resolve the myriad of stars into discrete images. The implication is, therefore, that improved optical quality resolves nebulous patches of the Milky Way into stars. In reaching his estimate of 10,000, Digges would need to pay special attention to faint stars, which are the most numerous and have the greatest effect on the total, and to stars in the crowded fields of the Milky Way.

Godfathers.

If Digges derived a quantitative estimate for the incidence of bright stars, did he try to quantify the incidence of fainter ones as well? Digges proclaims that the Universe is infinite, that the stars are numberless, that they are located at varying distances from the sun and extend through infinite space, but how could he justify the innumerate leap of faith from 10,000 to "infinity"?

Eyes and telescopes share the property that they are limited in both light gathering power and acuity. No star is infinitely luminous, so if they are scattered through space then it follows that, for any given telescope, there is always a limiting distance beyond which stars of a given luminosity go undetected. Digges credits Palingenius who saw, in his mind's eye, that the apparent brightness of stars could diminish owing either to increased distance, or diminished intrinsic brightness, or both acting together. Palingenius refrained from asserting a physical infinity, yet Digges unequivocally proclaims one. Digges could not *prove* the existence of a stellar infinity no matter how refined his work because detectors are never infinitely sensitive, yet both he and Shakespeare have no difficulty pronouncing that it exists.

The question devolves into how to read the book of nature through a telescope. In *Love's Labour's Lost*, Shakespeare opines that those who study in an arrogant ("saucy") and superficial manner are philosophically in error:

> These earthly godfathers of heaven's lights,
> That give a name to every fixed star,
> Have no more profit of their shining nights
> Than those that walk and know not what they are.
> Too much to know is to know naught but fame;
> And every godfather can give a name.

The Bard satirizes the stamp collectors of the sky who merely catalog and name celestial objects, as Hipparchus, Ptolemy, Tycho, Galileo and countless others since, have done. Merely listing objects hardly advances knowledge unless accompanied by analysis and interpretation as done, for example, by Schmidt and Green in their study of quasars.

In like vein, Digges, having counted the visible stars, may have taken the next logical step and counted stars up to the next threshold, which is that imposed by his telescope. It is not at all obvious that he could translate the numbers of stars he could see into an equivalent spatial distribution unless he performed experiments to correlate physiological sensation with physical stimulus. This seems unlikely and, even if he had done so, he probably could not have mounted arguments with sufficient clarity to convince skeptics of the significance of his observations, let alone add a mind-numbing and potentially unsustainable extrapolation to a physical infinity. Probably, Digges' positive assertion of a physically infinite stellar distribution is not based solely on empirical grounds.

Theological Infinity.

The argument hinges on the meaning of "infinity," which is a word that originated in theology. From 1413, "infinite" has the chiefly theological property of "having no limit or end ... either real or assignable ... boundless, unlimited, endless." This is the meaning intended by Palingenius and Nicholas of Cusa. It entered the language of mathematics through Euclidean geometry only in 1660 before acquiring its current spatial and temporal senses (*OED*). However, from about 1385, there existed "loose or hyperbolic" meanings as in "indefinitely or exceedingly great," "exceeding measurement or calculation," "immense," "vast," exceedingly and incalculably large as in "infynyte graciousnesse," "infinite nomber [of refugees]," or "goodis infinite." Shakespeare uses "infinite" in this loose sense when he writes in *Hamlet* of "virtues ... infinite," "infinite in faculty" and "infinite jest," and, in *Merchant of Venice*, of an "infinite deale of nothing." Three of the *Hamlet* "infinities" are hyperbolic, and it is reasonable to expect that the fourth, "infinite space," is as well.

Digges, too, uses "infinite" to describe the incidence of things that are arguably finite such as "infinite absurdities" and "infinite multitude of absurd imaginations," for no matter how many absurdities or errors this or another book contains, one must believe that the number is still arithmetically finite. An arithmetician realizes that he can conceive of a number that is so large that, in effect, he can call it "infinite" even though, of course, it is finite in the modern mathematical sense. We conclude then, that Digges means "infinity" to connote a limit that is exceedingly great but uncertain or incalculable. According to the Diggesian interpretation, therefore, an "infinite" distribution of stars describes physical existents that are accessible to the telescopically aided eye, beyond which they exist only in the mind's eye, and, as such, have an unobservable and incalculable bound.

Some argue that it was Copernicus who infinitized the Universe, but Copernicus tells us explicitly that he "leaves to the philosophers" the question of "whether the [W]orld is finite or infinite." Instead, he describes the outer boundary in finite terms as "the first and highest of all is the sphere of the fixed stars, which comprehends itself and all things" beyond which lies the "Artificer of all things" (Copernicus 8, 17, 24). The Diggesian "infinity" differs from the Copernican *immensum* in that Digges portrays a Universe that is inconceivably large, whereas the *immensum* is conceivably large because, by its Copernican definition, it has a definite bound. For Copernicus, the outer limit of the two-sphere Universe exists and the contrasting meaning of the Bard's "infinite space" favors the Diggesian interpretation.

A Perfit Description is short on details and long on infinity, suggesting that Digges mixes philosophical and religious principles with empirical data. He asserts that the stellar distribution extends infinitely up to the very court of the celestial angels and the "habitacle for the elect" and thereby inserts stars into a theological heaven. He includes God as an essential part of the quest to understand the natural World and criticizes those who rely strictly on their senses without proper regard for the role of reason, the latter being a gift from God "to lighten the darcknes of our vnderstandinge." Digges knows that humans have cognitive abilities in excess of those in animals whose primary activities are sleeping and eating, and advises us to put God's gift of brains to use in order to read the book of nature. The implied message is that humankind has the capacity to think abstractly and integrate physical theory with metaphysical belief.

By shattering the bounding Firmament, Digges shifts ignorance from the ineffable mysteries of the Empyrean to the indescribable outskirts of infinite space. Copernicus and his predecessors put a sharp division between physical and supernatural space, whereas Digges mixes the two. Many believe that, ultimately, the quest for understanding must resort to something supernatural, and that need would have been quite apparent in the sixteenth century. Fear of heresy hunters would motivate the shatterer to reach an accommodation between the two kinds of space and thus to design a new frame of creation that preserved both an incalculably large physical space and the divine abode.

Digges walked as much by faith as by sight because he faced the task of reconciling what he saw telescopically with what he believed theologically (cf. 2 *Cor.* 5:7). Nowadays, we know that as we look farther into space we look back in time and see the Universe in ever more youthful stages, so that the modern World view, with its seemingly intractable initial conditions and problematical end conditions, begins and ends with a question mark as well.

Visible and Invisible.

Hamlet accomplishes his mission through a relationship to the spiritual world. Hamlet does not know everything or foresee all events, for only the Artificer of all things is all knowing and all-powerful. Humans have free will, but many believe that there is a commanding otherworldly presence; events occur as if with the seeming happenstance of mundane life, yet adhere to a supernatural agenda. Hamlet and Horatio act, but, unassisted, their efforts are limited, so, when it comes to overthrowing corrupt regimes and replacing false ideology, the Supreme Power is not above reaching down to lend a helping hand.

Politically, Hamlet has toppled the Elsinore regime and paved the way for a new one. Corpses litter the path to literal success because in avenging the one death, eight others perish. The allegorical ending is not a tragedy, however, but a triumph. Hamlet dies when he has no further role to play, but his demise takes a back seat to the advancement of the New Philosophy. This is precisely the condition encountered in religious drama where "the real focus is not the Tragic Hero but the divine background" (Kitto 231).

In fact, some critics do see *Hamlet* as a religious play in which "Hamlet perpetually sees himself in a relationship with heaven and hell" (Edwards 40-50). Allegory is especially suited to this goal because, by the twelfth century, it had evolved into a universal vehicle of pious expression. The idea that *Hamlet* is a religious play finds support by evidence of supersensible influences that operate behind the scenes, guiding mortals toward a deeper understanding of nature and the New Philosophy. The challenge is to understand physical space as an amalgam of natural and supernatural existence. Thomas Digges freely admits that he is not up to the task. Evidently, it is incomprehensible to Shakespeare, too, for he, as a mere mortal in a milieu of limited knowledge, would not presume to pontificate on the theological and physical implications of the new World view, but prefers to leave these difficult questions to future generations of eyases with insight to match eyesight.

The revolution that Copernicus and Digges fomented occurs in a broad context of the triumph of good over evil and challenges human perceptions of theological infinity. *Hamlet* is anagogical, for its meaning transcends literal, allegorical, and moral senses to embrace a fourth and ultimately spiritual sense. In neo-Platonic and neo-Pythagorean traditions, the transcendent One, or the Good, is the source of all reality. Following Plotinus, the Eternal One created the Universe and, through intermediaries and the principle of emanation, continues to play a role in it. The Supreme Being, the Divine Intelligence after whom humans are fashioned, is Theos, akin to God in Judeo-Christian theology, and in *Hamlet,* this Distant Drummer compels a new resident of the spiritual world, the spirit of Old Hamlet, to cross the divide separating natural and supernatural space in order to lift humankind from the intellectual abyss of the Dark Ages.

Hamlet establishes and dramatizes the Hebraic concept of a covenant between God and humankind, in this case to transform the old perceptions of the World. The tragic hero is "in a dramatic plot overseen and guided by Providence" and "rebelling against the divine order of the Middle Ages" (Frisch 16-7). Hamlet must set matters right and, even though his life gives the appearance of free will, there exists a commanding influence that speaks through him and acts for the good. The spiritual image is of a beneficent One, like the Creator of Heaven and Earth and of all things visible and invisible.

Summary of Facts.

The foregoing discussions suggest that *Hamlet* announces for the first time certain empirical facts on the heavenly bodies. In summary, results for Solar System objects are:

> The Sun has dark spots on its surface that come and go.
> The Moon's surface is blemished.
> The Moon and Venus are not self-luminous.

Venus is resolved and has phases like the Moon.
Mars is resolved and has no phases.
Jupiter is resolved and has no phases.
Jupiter has a surface red spot that is resolved.
Jupiter rotates.

For stars, the results are:

The heavens are mutable (Olson, Olson and Doescher).
The New Star of 1572 erupted on November 3.
The Milky Way is comprised of stars.
The perspective glass does not resolve stars.
There are about 10,000 stars visible to the naked eye.
Stars are self-luminous like the Sun.
Stars are scattered through infinite space.

These conclusions rest upon the proposition that Thomas Digges used a telescopic magnifier. This gave him the confidence to risk printing novel interpretations at a comparatively early time and, similarly, inspired Shakespeare with confidence to write of them. The puzzle unravels further on considering the winds of change.

CHAPTER 11: WINDS OF CHANGE

If hereafter, anatomizing this surly humour, my hand slip, as an unskilful prentice I lance too deep, and cut through skin and all at unawares, make it smart, or cut awry, pardon a rude hand, an unskilful knife, 'tis a most difficult thing to keep an even tone ...

John Burton 1628

While in Augsburg in 1575, Tycho Brahe invited Tobias Gemperle (1550-1602) to visit Denmark. Gemperle accepted the offer and, while in residence, produced portraits of Danish nobility and pictures of their castles. He spent the years 1577-1587 on the island of Ven and produced an assortment of images of Tycho and his instruments. One shows Tycho with his great mural quadrant, with one hand pointing toward an opening in the south wall of the room that housed the quadrant and the other pointing to an open book resting on a nearby table.

Tycho turns Forty.

Gemperle was probably the artist who in December 1586 produced a sketch of Tycho on his fortieth birthday. Tycho liked the sketch and commissioned Europe's renowned engraver, Jacob de Gheyn II (1565-1629), to produce an engraving based on it (see Figure 7). Tycho may have selected de Gheyn not only for his artistic skill but also because of his latent interest in astronomy, which, in 1600, led to nine splendid woodcuts depicting constellations mentioned by the Roman poet, Caius Julius Hyginus (1st century AD).

De Gheyn's engraving chronicles Tycho's accomplishments up to age forty. It shows him as a sort of "celestial lord" (Christianson 114) framed by an architectural motif comprising two pilasters supporting a semi-circular arch. The structure holds the heraldic shields of his sixteen great-great-grandparents, with four distaff escutcheons attached to each of the two pilasters and the eight spear shields affixed to the arch.

Figure 7: First engraving of Tycho Brahe by Jacob de Gheyn II, 1586.
(Huizinga, from Strunk 74).

Figure 8: Second engraving of Tycho Brahe by Jacob de Gheyn II, 1586.
(From *Epistolarum astronomicarum*.)

Although the general design was common at the time, it may have suited the "celestial lord" to see himself perched at the interface of Heaven and Earth, with the Earth below and the arch above signifying the shell of stars. The lofty status of the spear escutcheons symbolizes the stellar pedigree of his male ancestors. The female shields are affixed to the two pilasters that rest on the corrupt Earth and the male shields flying high in the heavens where perfection rules, so one might wonder whether the fair sex is less favorably situated than the male, but note that it is the female architecture that props up the male.

Jacques de Gheyn signed the engraving, although some copies also carry the legend *Marco Sadeler excudit* ("struck out by Marcus Sadeler"). A de Gheyn drawing undoubtedly preceded the engraving but is lost. In the engraving as in the woodcut preceding it, Tycho appears from the waist up in three-quarter profile, almost en face, with a greatcoat draped over his shoulders. His gold chains and medallions are much in evidence. A plumed beret rests next to his right hand on top of the parapet. His left hand is above the breastwork clasping a glove, the conventional symbol of gentility. His torso turned slightly to his left. The lighting comes from Tycho's right, correctly illuminating both his right side and the corresponding reveals (facets, jambs) of the framing architecture.

Errors.

The engraving had two genealogical errors, so Tycho commissioned a second engraving (see Figure 8) in which "Rønnor" and "Stormvase" replaced "Hønnor" and "Belker." There is a general resemblance between the two, but the differences are noteworthy. Tycho has donned his plumed beret, *Instrumentorumque* replaces *Machinarumque*, there is a mirror reversal of his torso but not his head and his facial features are redrawn. The mirror reversal has Tycho's left hand resting atop the parapet and his right hand clutching the glove, instead of the reverse, and details like arms, chains, and medallions are not precisely the same. Gemperle's original drawing shows signs of having been used for tracing, which may explain the partial mirror reversal.

In correcting these errors, de Gheyn added new ones. Light falling on Tycho's face and torso comes from Tycho's left instead of his right, but the light that illuminates the structure still comes from his right, as in the first engraving. Conceivably, two streams of light could achieve the shadowing as depicted if beamed precisely in such a way as to illuminate Tycho from one side and the architecture on the other, but this explanation is utterly contrived. The question arises, why would an engraver of de Gheyn's stature commit such elementary errors when the needed corrections involved only two names and a word on the plaque?

In Gemperle's original drawing, Tycho's right hand rests with palm facing downward on a flat surface. Gemperle depicted Tycho's right hand with what appears as a deformed "little" (i.e., fifth) finger, whereupon de Gheyn faithfully reproduces the feature in both engravings. Normally, when a hand rests on a flat surface, the thumb is the only finger that can bend more or less in the plane of the surface, but in all three works, the little finger has this property. It looks as if it might have been broken at its knuckle and twisted through about 90°, causing it to resemble a thumb.

A little finger cannot bend as depicted unless a result of malformation, misadventure, or illness. Such a deformity might occur in someone who suffered from arthritis, but a limited search of the medical literature failed to turn up anything resembling the case. Polio is a possibility and, in fact, in the twentieth century, the illness did affect a finger of the astronomer Bart J. Bok (1906-1983) (Levy *Man* 42). There is no such record in Tycho's case, however. Owing to the mirror reversal, one could argue that both of Tycho's little fingers were identically deformed, which is unlikely. Even if arthritis or polio were the culprits, historians would surely have noted such a disability in one who was a prolific instrument builder and proficient observer. No deformity was recorded when his remains were exhumed in 1901, nor was there any expectation of any (Matiegka). Many portraits and bas-reliefs of Tycho survive to this day and all have normal hands. Tycho's other hand (the one without a deformity, sinister in Figure 7, dexter in Figure 8) appears to grasp gloves awkwardly. The depiction looks unnatural and not up to the standards of Gemperle and de Gheyn. The thumb of that other hand is unseen, but perhaps it is vertical, out of sight behind the gloves. Yet the little finger of that glove-grasping hand is somewhat pronounced in the

Gemperle sketch and even more pronounced in the engravings, raising further questions.

Answers.

The following description refers to Figure 7 and words in parentheses refer to the Figure 8. The little finger of Tycho's resting right (left) hand looks like a thumb. No fifth digit of that hand is visible because it lies hidden behind the rest of the hand. As a result, Tycho's right (left) hand resembles a left (right) hand whose palm and middle three fingers hide the little finger. The awkwardness with which Tycho grasps the gloves in his left (right) hand is due, in part, to the absence of a visible thumb on that hand. Only the other four digits of that hand are visible as they curl around the gloves. A simple explanation is that Tycho holds the thumb of his left (right) vertically, and thus out of sight, behind the part of the glove that protrudes above his fist. Tycho could have tucked his thumb behind his other four fingers and the glove, as one might do when "holding thumbs," but the possibility seems unlikely.

It appears as if both artists switched Tycho's hands left and right. To test the idea, we must examine the lack of a thumb on the glove-grasping left (right) hand. In both engravings, all we see are four fingers wrapped around the gloves, so the left (right) hand could just as well be a right (left) hand, with the back of the hand facing forward. Then, since the apparent glove-grasping left (right) hand is, in reality, a right (left) hand, the thumb of that hand would point downward, in the same direction as the fingers of the gloves. In order for the thumb of the glove-grasping hand not to show, the dangling fingers of the gloves would have to cover it completely. Inspection of the first engraving allows for this possibility because two of the fingers of the glove rest on the parapet without any spaces between them and could easily obscure a downward-pointing thumb. The second engraving obscures the thumb even more convincingly because the fingers of the glove fall behind the parapet where there is more shadow and where a downward-pointing thumb could more easily lurk unseen. By swapping Tycho's hands in the second engraving, de Gheyn depicts Tycho giving himself a "thumbs down" with both hands.

Gemperle was capable of drawing and painting hands properly. In 1578, he painted a portrait whose face and hands were so realistic that they captivated the viewer. In the following year, he made the first medical illustration in a Danish book – of a human skeleton no less. Available evidence indicates that Gemperle was an able artist who was well versed in anatomy. De Gheyn, too, was renowned for precise and truthful likenesses. If Tycho really did have such a deformity, why did both artists portray it when, ordinarily, artists seek to gloss over the physical defects of their subjects? Part of the answer lies in the fact that Gemperle was capable of painting symbolically and allegorically, an ability that, ironically, had appealed to Tycho.

The simplest explanation is that Gemperle and de Gheyn deliberately created and perpetuated the malformations. Why, then, did Tycho not object from the outset? Was he not paying attention, or was his aesthetic appreciation so impaired that he was blind to anatomic abnormality? Did the portrayal of his medals so enrapture him that he paid no attention to other details? Did he focus attention more on depictions of his nose? Tycho had nothing but praise for Gemperle and de Gheyn and, if he did notice the anomalies, did he attribute them to artistic whimsy and ignore them simply because of the reputations of the artists? What motivated the prank in the first place?

Motive.

Gemperle had plenty of opportunity to get to know Tycho and he would have discovered what is today generally accepted, that Tycho was egocentric, ill tempered, autocratic, quarrelsome, and domineering. He was a scholar but hardly a student of humanity. Some people, the local villagers in particular, hated him because of his authoritarian ways, but they dared not express their anger openly. By the end of 1596, Tycho had worn out his welcome and, in the following year, having engendered the ire of the new Danish monarch, King Christian IV (1577-1648), he left Denmark. By then, Gemperle and de Gheyn had had their say. Gemperle used art to express his opinion of Tycho's quarrelsome personality, and de Gheyn, perhaps out of respect for a fellow artist, went along with the prank.

199

Artistic Tools.

Did anyone else understand conditions at Ven well enough to see through the Gemperle-de Gheyn japery? A good candidate would be the poet to whom Thomas Savile forwarded Tycho's engraved portrait and request for witty epigrams and, as noted, a likely candidate for that honor is Shakespeare. Savile would know that a foreign celebrity like the esteemed astronomer Brahe deserves nothing but the best. Both Savile and the poet would consult with Digges because not only was Tycho an astronomer who, like Digges, had produced a book on SN 1572, but because evidence indicates that Shakespeare knew the Digges family.

Thomas Digges likened the Ptolemaic model to a human frame, with head, hands, and feet taken off different men and cobbled together. Bits and pieces comprise both the image of Tycho and his hybrid model as well. An imaginative poet might follow the lead of Digges, Gemperle and de Gheyn and use poetic tools to amputate Tycho's hands and reattach each in place of the other. If the poet is not too fussy about medical procedures, he might have recourse to the tools of a carpenter, among them a handsaw. Reference to this tool occurs in Hamlet's much celebrated yet perennially baffling passage in 2.2: "I am but mad north-north-west. When the wind is southerly, I know a hawk from a handsaw." Hamlet speaks of weather vane directions, a bird, a tool, and his alleged madness. The passage is baffling because the two directions are not quite opposite one another and birds and tools have little in common.

Tycho was a pioneer in cartographic measurement who, in 1596, produced what is generally regarded as the world's first map based on triangulation. The map was of Ven whose location he determined relative to the surrounding shores of the Sound. The origin of coordinates at Ven is consistent with the fact that Tycho regards his island as supremely important. In *Progymnasmata,* for example, *A Catalogue of some of the most eminent Cities and Towns in Europe with their Latitude, and difference in meridians from London,* includes Uraniborg but not Wittenberg! This occurs despite the fact that a full-length figure of Copernicus decorated his large wooden equatorial armillary, built about 1580. Claudius Ptolemy is depicted as well, along with the Arab prince Al-Battani (*c.*858-929) who tested many of Ptolemy's results with fresh observations. Also depicted is the vain surveyor, Tycho, himself.

To a good approximation, geodetic lines proceeding from Tycho's observatory on Ven in directions NNW and S pass through Helsingør and Wittenberg. Thus, NNW points more-or-less to the bastion of geocentricism where Hamlet must feign madness and S points to the place where heliocentricism flourished and where Hamlet has no need of an antic disposition.

Hamlet can distinguish between a correct and an incorrect model of the World just as he can tell a hawk from a handsaw. However, this does not say much of Hamlet's powers of observation, for surely no one could confuse a bird and a carpenter's tool. This difficulty has prompted a hubbub of conjecture on the meaning of the word "handsaw," not to mention the role of Hamlet's alleged madness.

Comparisons.

When Shakespeare uses a word, he often has all possible meanings in mind. In the passage in question, suppose that Shakespeare compares comparables; i.e., either "handsaw" and "hawk" are both tools, or they are both birds. Only after 1700, does "hawk," mean a tool used by plasterers (*OED*). Unless the *OED* does not record earlier usage, it is unlikely that "hawk" and "handsaw" are both tools. An alternative is that "handsaw" refers to a bird, and some argue that the word, suitably slurred, means "heron," "hernshaw," "heronshaw," or even "heron pshaw!" True, this contrast is between different kinds of the same genre, but hawks and herons are so different that a capacity to distinguish between them does not seem a particularly noteworthy accomplishment for a prince who, one expects, is well versed in falconry. It appears that "hawk" and "handsaw" are neither both "tools" nor both "birds." Perhaps, instead, they have some other feature in common. As noted, "handsaw" is reasonably a reference to Tycho's artistically swapped hands, but it makes little sense for Shakespeare to associate a bird with Tycho's depicted malformation, so perhaps there is something about the bird in question that we do not understand.

The direction of south points from Tycho's scientific home to Copernicus' scientific home. Hamlet is well equipped to distinguish between the corresponding geocentric and heliocentric models, but what needs explaining is why he picks a "hawk" to contrast with a "handsaw."

A "hawk" is a bird of prey renowned for its remarkable eyesight, and Hamlet would know that the hawk's ability to spot prey far outstrips that of the unaided human eye. By contrast, the resolving power of the human eye is 30 to 60 arc seconds, at best. Tycho is the astronomical observer who made naked eye observations and is associated with the crude surgical tool, a "handsaw," so perhaps the Bard contrasts Tycho's visual acuity to that of a hawk. Surely, parallelism demands that, if "handsaw" refers to one famous observer of the skies, then "hawk" refers to another.

From the eighth to the nineteenth centuries, the word "hawk" meant, "any diurnal bird of prey used in falconry," of which there are many (*OED*). From 1550 to the eighteenth century, hawks included sub-species "leonardes," "leonerettes," and "fawcons." Thus, as "handsaw" conceals a reference to Tycho Brahe, so "hawk" conceals a reference to a *leonard*, or, perhaps, to Leonard Digges. As far as visual acuity is concerned, Leonard Digges and Tycho Brahe are as dissimilar as a hawk and a handsaw. Leonard invented the first two-element magnifier, which increased visual acuity to some value in the range 1-3 arc seconds, whereas Tycho built instruments suitable for naked-eye observations with acuity about 30 times worse. In other words, Leonard the "hawk" has great visual acuity thanks to the perspective glass, whereas the artistic "handsaw" refers to the observer with severed and reconnected hands whose acuity is comparatively poor. In the second engraving, *Instrumentorum* replaces *Machinarum* and, correspondingly, Tycho's naked eye "instruments" become the butt of the hawk vs. handsaw comparison.

The two wind directions point toward the two prevailing influences on Tycho's World model. Shakespeare speaks approvingly of the heliocentric southerly influence from Wittenberg and disparagingly of the north-north-westerly geocentric influence from Elsinore. The hybridized geocentric model enjoys a southerly influence to the extent that it upholds Tycho's choice of the Sun as the center of the orbits of the five unresolved Ancient Planets, but the influence emanating from the NNW is undesirable because Elsinore champions the centricity of the Earth. Tables 4a and 4b give a summary of results.

Table 4

SUMMARY OF MULTIPLE MEANINGS, *HAMLET* 2.2.347-8.

(a) Hamlet's Dispositions and Tycho's Model Components.

DIRECTION	SITE	MODEL	DISPOSITION
North-North-West	Elsinore	geocentric	mad
Southerly	Wittenberg	heliocentric	discerning

(b) Hamlet's Discernments when the Wind is Southerly.

CATEGORY	SCIENTIST	OPTICS
Hawk	Leonard Digges	telescopic
Handsaw	Tycho Brahe	naked eye

The Tragedians.

Shakespeare verifies, in several ways, that Hamlet's hawk and handsaw remark concerns Tycho and Thomas Digges. To begin with, a mere thirteen lines before it, Rosencrantz dubs child actors "little eyases" nestling in an "eyrie," meaning that they are fledgling hawks. The context is the arrival of adult thespians that are on their way to cheer up the prince who suffers madness in a chilly wind. Hamlet quizzes the courtiers on the nature and reputation of the new arrivals, and Rosencrantz explains that the fad of child actors forced them to seek work on the road. Before the War of the Theatres, these touring thespians were content to see Hercules standing outside the Globe Theater and bearing the weight of the world upon his shoulders. The myth is that Atlas, the first astronomer, holds the sky on his back, except for one occasion when Hercules relieves him. Hercules had once rescued Atlas' daughters, and Atlas was so grateful that he not only gave Hercules the honor of holding up the sky for a while but also taught him astronomy. Gerardus Mercator (1512-1594), a Flemish cartographer for whom the Mercator projection of 1568 is named, thought Atlas supported the Earth, so he called his collection of maps an "Atlas. Withal, Atlas is frozen in place as becomes the bearer of a stationary burden and, when the Titan does take a break, his stand-in is statuesque as well. Hercules and his load are emblematic of the geostatism of the Old Astronomy under whose aegis the thespians were once content to play.

However, a certain "innovation" has cost them their jobs. "Innovation" means, primarily, "revolution," as used in 1596 in *Henry IV part 1*. After 1553, it meant "the alteration of what is established by the introduction of new elements or forms" (*OED*). Hamlet's "transformation" and "revolution" fit the definition of "innovation" and the meanings apply, reasonably, to the new World view. A second meaning of "innovation" is a "political revolution ... rebellion or insurrection." The revolution that Shakespeare has in mind applies as well to the political scene at Elsinore and the theaters, as to the revolution in World view that Hamlet later addresses.

Hamlet wonders whether the thespians are superannuated but Rosencrantz assures him that, "their endeavour keeps in the wonted pace." In the sixteenth and seventeenth centuries, "pace" also meant a company or herd of asses (*OED*), so Rosencrantz inadvertently associates the hirelings with creatures renowned for slowness of wit. The irony is that Rosencrantz, associated with the naked-eye astronomer Tycho and his model, announces with remarkable foresight that fledgling successors to Leonard, the "hawk," oust naked-eye observers who practice the old ways. These young eyases are "tyrannically clapped" for their performances, anticipating that future telescopists will play to great acclaim. Perhaps Shakespeare warns, too, that science, for all its high-mindedness, is not immune to tyranny.

The flow of the dialogue establishes that the touring thespians no longer enjoy the same esteem as they did when they played in the city. Hamlet does not find this surprising. "It is not very strange," he counters because he knows that others suspect that geocentricism is passé. Hamlet continues, "for my uncle is king of Denmark, and those that would make mouths at him while my father lived give twenty, forty, fifty, a hundred ducats apiece for his picture in little." Before Claudius became king, no one could care less about him, but now that he is the chief defender of the status quo, everyone wants a copy of his picture. Portraits of Ptolemy, imaginative and otherwise, have accumulated over the years. In the sixteenth century, several relatively small portraits were produced that may have commanded high prices, from twenty to a hundred ducats perhaps, which is a measure of the esteem of Claudius Ptolemy and geocentricism in the sixteenth century.

Tycho's Garb.

The oath "'Sblood" is a contraction of "his blood" and is thought to refer to the blood of Christ (Andrews 2.2.393n). This oath, however, comes straight after Ptolemy's "picture in little," suggesting that Shakespeare switches suddenly from a pure to a hybrid geocentricist, in which case the contraction refers to blood associated with Tycho's little picture. There are two sources of Tychonic blood, and Gemperle and de Gheyn depict each of them. Tycho bled when he lost his nose, and de Gheyn imagined that Tycho bled when he amputated and reconnected his hands. Disjointed hands and a prosthetic nose are "more than natural," as is the false lighting of the second engraving, which is a "picture in little." Then, when welcoming the adult players, Hamlet suddenly refers to "hands" supposedly because good actors deserve applause (Jenkins 2.2.366-9n), but the remark could also refer to Tycho's anatomical confusion. The next sentence reads, "Th'appurtenance of welcome is fashion," in which the word "fashion" may refer to the extravagant garb depicted in the engravings. Then (wonderful to relate) the word "garb" occurs in the next line, "Let me comply with you in this garb." "Garb" means stylishness of appearance, or more simply "fashion" but may refer also to a "manner of doing things" (*OED*), such as Tycho's manner of observing or the manner of actors doing things on stage. Tycho's portrait also shows appurtenances that complement his garb, such as a plumed bonnet, gold chain, and medallion. After the Players leave and re-enter at the start of 3.2, there is more hand sawing as Hamlet instructs them, "Nor do not saw the air too much with your hand thus." The mention of "air" in this passage sustains the imagery of Tycho and his false nose. Within a few lines, Shakespeare alludes directly to Tycho's childhood abduction by his uncle and aunt, employing the words "uncle-father" and "aunt-mother." All these allusions occur in the context of "welcoming" the adult players who are devotees of the weight traditionally resting on the shoulders of that illustrious Titan, Atlas, whose load is sometimes the sky with its hard, opaque shell like that of a nut.

Another reference to Tycho's garb occurs in 5.2 when the royal minion, Osric, enters. Hamlet instructs him, "Put your bonnet to his right use, 'tis for the head." This is exactly what the engraver de Gheyn did in the second engraving. The Osric dialogue commences with confusion about whether the weather is hot or cold, as befits how Tycho wears his greatcoat, which, in both engravings, he only drapes over his shoulders. Osric tries to justify doffing his bonnet by complaining, "it is very hot." In the Northern Hemisphere, hot weather often results when breezes are southerly. Osric is trying to pull the wool over Hamlet's eyes by making it seem as if he is a southern, Wittenberg sort of person, but Hamlet knows better as he responds, "No believe me, 'tis very cold, the wind is northerly." Winds coming from a northerly direction are seldom warm and Hamlet has already associated them with the ill wind that blows Tycho's hybrid model no good. Hamlet corrects Osric by denying that the messenger brings with him any hint of milder weather. Osric humors Hamlet by agreeing with him. "It is indifferent cold," he says, whereupon the mad Prince does an about-face and remarks, "But yet methinks it is very sultry and hot for *my* complexion" (emphasis added) which identifies Hamlet's cosmic being with the southern direction. These contradictions flummox Osric who humors the Prince once again by agreeing with him. Before Osric gets to announce the king's wager, therefore, Hamlet has announced that he knows which way the wind blows, and it blows both hot and cold.

Both the hawk and the southerly wind direction associate the inventor of the optical magnifier with the intellectual home of heliocentricism whose instigator was, primarily, a theoretician and not an experimentalist. This suggests that Shakespeare supports Copernicus not simply for his novel theory of the World but also because empirical evidence exists that favors it. Like many scientific advances in recent memory, instrumental innovation and new technology drove the revolution in World view, but the mere existence of new technology is insufficient to foment revolution because scientific conclusions require the intelligent application of scientific methodology. Shakespeare's sub-text concerns the grounds of knowledge and establishes *Hamlet* as an account of the rise of the New Philosophy, as we see next.

CHAPTER 12: LOVE LETTER

Tragedy is a representation of an action that is worth serious attention.

Aristotle

Hamlet's love letter contains a verse that deserves scrutiny if only because its first two lines refer to two basic tenets of the medieval World view, that the element Fire accounts for the light of the stars, and that the Sun is in daily motion around the Earth. The letter is all the more noteworthy because Shakespeare introduces the idea of Hamlet's "transformation" at the start of the selfsame scene that contains many other allusions to astronomy. The letter and its contents are notoriously puzzling.

Contacts.

In 1.3, Polonius forbids Ophelia to have further contact with Hamlet, but, in 2.1, Ophelia complains that Hamlet has frightened her. Polonius asks her whether Hamlet is mad for her love. She says that she does not know, but assures her father that she repelled his letters and denied his access to her. Polonius immediately leaps to the unfounded conclusion that Hamlet is mad. For some reason, Polonius has a great need to derail his daughter's budding relationship with the nominal heir to the throne. He must avoid confronting Hamlet directly, so he vows on two occasions to take Ophelia and report the incident to the king. When he enters in the next scene, however, he enters alone, albeit armed with a letter that, he says, Hamlet wrote to his daughter. Perhaps Hamlet wrote and delivered this letter before Ophelia promised to repel subsequent letters (Wilson *Happens* 113) and then, after Hamlet frightened her, she may have given it to her father. If so, Polonius has no need for testimony from his daughter, for the letter proves his theory exactly.

Missive.

Polonius reads the letter and censors part of it ("et cetera"):

207

To the celestial, and my soul's idol, the most
beautified Ophelia, ... In her excellent white
bosom, these, et cetera ...

Doubt thou the stars are fire,
Doubt that the Sun doth move.
Doubt truth to be a liar,
But never doubt I love.

O dear Ophelia, I am ill at these numbers, I
have not art to reckon my groans; but that I love
thee best, O most best, believe it. Adieu. Thine
evermore, most dear lady, whilst this machine is
to him, Hamlet.

The letter is "a great puzzle" and the formulaic quatrain is "affected, juvenile, and graceless" (Edwards 2.2.109-22n).

Polonius has not even finished reading the first sentence before Gertrude asks in disbelief, "Came this from Hamlet to her?" We, too, wonder why Hamlet would burden his inamorata with such inanities. Whereas the stars serve the interests of romance quite well, the Moon provides the desired ambience, but Hamlet is so gauche that he ignores it in favor of the Sun. By calling Ophelia "celestial," the earnest swain oversteps his bounds because it is ridiculous, even blasphemous, to think that any terrestrial existent can aspire to celestial perfection (excluding of course the divine gift of the human soul, which is not at issue here). The double superlative, "I love thee best, O most best, believe it," emphasizes the extremity of his devotion. His letter abounds with adjectives and his grammatical constructions are convoluted, all of which are grist to the pedant's mill for it appears that Hamlet is much enamored of Ophelia but cannot write from the heart. It seems that Hamlet is in love, but mad.

Doggerel.

For ease of reference, let the four lines of the verse quoted above have labels L1, L2, L3, and L4. L1 and L2 both concern the Old Astronomy. L1 refers to the property that both fixed and wandering stars shine by virtue of the element fire, and L2 refers to the Sun moving around the

Earth. L3 is a trite observation on truth and falsity and L4 says that Hamlet loves Ophelia. The word "doubt" begins each of L1-3 and recurs in L4. It means "to be uncertain or divided in opinion about" or "to call into question" (*OED*) and, in urging examination of the letter, Sohmer (private communication) stresses its repetition.

L1 and L2 urge Ophelia to doubt two beliefs of the standard cosmological model. Commentators have noticed that the lines suggest acknowledgement of the New Astronomy, but interpretation of L1,2 requires consideration of the broader context of L3,4 and the rest of the letter, as well as the rest of the play.

In L3, there is a change of subject. If we believe that "truth" refers to the totality of real things, events, and facts, then L3 means, in effect, that Ophelia should doubt that "a liar" is associated with the grand ensemble of all knowledge and lore. No right-thinking believer in received wisdom should doubt that truth is truth, and everyone should doubt that truth is a "liar."

There is a lack of parallelism between L1,2 and L3. In a standard interpretation, L1,2 concern a World view that nobody in his or her right mind should doubt, but everyone should doubt that truth is a liar. In the sixteenth century, received wisdom was sacrosanct; the Universe was an orderly and hierarchical construct in which there were answers for all questions and no room for ambiguity or doubt.

L4 resolves the orthogonality of L1,2 and L3. It begins with the word "but" that often introduces a statement of the nature of a contrast to what has gone before (*OED*); i.e., it connotes exception. However, "but" can also express disconnection (*OED*), in which case it has a closely equivalent meaning of "moreover" (*OED*), although "and" would do as well. Thus, "but" can introduce a distinct or independent fact. We suppose that, when allied with L1,2, L4 expresses exception, and when allied with L3, it introduces an independent fact.

Consider the combination L1,2+4. "Doubt" in L1,2 has the "weakened sense" of "suspect" (*OED*), similar to its occurrence some fifty lines earlier where Gertrude suspects the causes of Hamlet's disposition, although, of course, she does not actually know them. Hamlet uses "doubt" rhetorically to urge Ophelia to "mistrust," or hesitate to "believe," or hesitate to "trust" (*OED*). Hamlet tells her to doubt that which no sane person would doubt, before she doubts his love for her. In this way, he tells Ophelia that he loves her absolutely because perfection resides only in the heavens. This explains Hamlet's epithet "celestial" with which Hamlet compliments Ophelia. Thus, L1,2+4 mean that Ophelia may go so far as to doubt two sacred verities of contemporary cosmology BUT she must not doubt Hamlet's love for her.

In L3, "doubt" pertains to something that is true in its own right because truth is hardly a "liar." In that case, "but" in L4 has the meaning of disconnection and serves to introduce a distinct or independent fact as might occur in the minor premise of a syllogism (*OED*). Thus, the combination L3+4 contains two "distinct or independent" facts, that Ophelia should doubt something that is plainly false AND she must not doubt Hamlet's love for her.

L1,2 and L4 are contingent upon one another whereas L3 and L4 are independent. Edwards (2.2.117n) writes that each of L1-3 "means the same." Ophelia may challenge the unchallengeable, but, no matter whether she doubts things that are true or false, she should not doubt Hamlet's love for her. With due respect for the perils of reductionism, the literal meaning of the entire verse in this standard interpretation is simply that, above all, Ophelia must not doubt that Hamlet loves her.

The question remains: why does Hamlet write a love letter in such an odd way? Some people appreciate and recite poetry yet cannot write it themselves, but there is no evidence in the script or in Shakespeare's sources that says that unrequited love engenders illiteracy. Other puzzles exist, like Hamlet's emphasis on matters celestial and the sudden and confusing shift in meaning from L1,2 to L3. Moreover, the meaning above pertains only to a literal interpretation of the play and it behooves us to seek allegorical understanding as well. Let us take the verse line by line.

Fire Revisited.

The concept of Fire requires re-examination since L1 urges its rejection. Chapter 1 describes how Copernicus assigned the Sun and stars to a state of rest, which means that the old theory cannot account for their luminosity and, since Moon and Earth are not self-luminous either, the only candidates left that might shine by Fire are Mercury, Venus, Mars, Jupiter and Saturn. From Chapter 10 we see that Venus shows a full range of phases like the Moon and that these phases correlate with the planet's heliocentric orbital position relative to the Earth, making it a dark star like the Moon.

In *A Perfit Description*, Digges says that planets might glow if suffused by rays of sunlight ("persed with solar beams"). Having absorbed sunlight, they would re-emit in all directions and act, effectively, as scatterers of sunlight. However, if the Sun does not shine by Fire, then these five supposed light-scattering objects do not either, even indirectly. Consequently, the number of planets where Fire may burn is reduced to four. Reasonable people would conclude that Fire is dead and literature bears them out.

Donne's Criterion.

In 1610-1, Galileo's announcement that Venus has phases like the Moon sent shock waves through the literate world, moving John Donne to put pen to paper in service to the New Astronomy. The following year, Donne stated that the New Philosophy "calls all in doubt" because it "arrests the Sun," "bids the passive earth about it run" and puts out Fire (see Chapter 2). Donne was familiar with the work of Copernicus and Galileo and it is no coincidence that he penned these lines one year after Galileo had published his telescopic results. Here at least, Donne doubts the Old Astronomy and it is again no coincidence that, a decade earlier, Shakespeare had made "doubt" the basis of Hamlet's doggerel.

In 1611, only a few people knew of the telescopic results that Shakespeare had reported in *Hamlet,* so Donne's poetry provides an objective criterion by which to judge heliocentricism and the end of Fire. To Donne, Galileo's discovery that Venus was a dark star tipped the scales in favor of Copernicus and heliocentricism. Let us call this realization, "Donne's criterion." Applying it retrospectively to *Hamlet,* we must conclude that the telescopic results on Venus reported by Shakespeare allow him to draw the same conclusion that Donne drew.

Shakespeare, in 1601, and Donne, in 1610, reach the same conclusion that Fire is put out. The only difference between Donne's and Shakespeare's pronouncements is that Galilean data emboldened Donne to speak out, whereas Shakespeare masked his conclusions. It

211

is possible, therefore, that the Bard intends L1 to mean simply what it says and the parallelism of L1 and L2 suggests also that he intends L2 to mean what it says too. If so, two possibilities need exploring.

One Alternative.

In L1, Hamlet advises Ophelia to doubt that the stars are fire. By the precepts of the New Philosophy, this is correct. In L2, he advises her to doubt that the Sun is in motion. Correct again! L1,2+4 then mean that Ophelia must doubt the Old Astronomy BUT she must not doubt Hamlet's love for her. In L3,4, Hamlet changes both the subject and the meaning of "but," as discussed above. In L3, he advises that Ophelia should doubt that truth is a liar, which seems fair enough. In L4, he tells Ophelia that he loves her. There is no surprise there, either. Then, L3,4 combine to mean that Ophelia must doubt that truth is a liar AND she must not doubt Hamlet's love for her.

Concerning the first combination L1,2+4, mythology justifies Hamlet's epithet "celestial" because, to the ancient Greeks, Phoebus Apollo was god of the Sun and Diana was goddess of chastity and of the Moon. If she were to team up with Hamlet and forsake her father and his primitive worldview, unbounded heliocentricism could prevail at Elsinore. Hamlet's admonitions, L1,2, are not merely rhetorical as in the standard interpretation, because, in conjunction with L4, Hamlet advises Ophelia that she must prepare to help rule Elsinore just as, in mythology, the royal Sun and the chaste Moon rule the sky.

In the new Copernican scheme, the Moon is the only Ancient Planet to retain a geocentric orbit. This implies that the lunar Ophelia has no inherent reason to doubt that the Earth is the center of her ambit. In L2, however, Hamlet states the primacy of the Sun and apprises her of her potentially new and unique role. She retains geocentricity to the extent that she will forever be the child of a geocentric father, but her new orbit would expand her horizon and carry her around Hamlet's Sun. The prospect of this liaison seems delightful, except that, when the potential alignments are most nearly exact, either an eclipse of the Sun or the Moon occurs. Hamlet and Ophelia both suffer eclipse, Ophelia being too much of the watery Moon and Hamlet too much in the Sun. Their demise does not occur simultaneously, and neither do eclipses of Sun and Moon.

Hamlet must know that Ophelia's name derives from the celestial alignments when the Moon is Full or New, since it refers both to the alignments of Lunar Opposition and Conjunction with the Sun. Thus, Hamlet speaks of his alignment with Ophelia, which, in the abstract world of allegory, is that of Syzygy, but which, in the literal interpretation, passes equivocally as "love." The heir-apparent needs his cosmic alignments just as the false monarch needs his conjunction with Gertrude. Hamlet is the heir-apparent to the throne and is a prince out of Ophelia's star because he represents the star that rules the planetary system, whereas Ophelia represents merely a satellite of a satellite of that star and so moves in decidedly inferior circles. Ophelia's star rules Neptune's Empire whose high tides are deepest at times of eclipse, at which time the ebb and flow of events are most likely to swamp the unwary. After Hamlet forsakes her, she is out of her depth and perishes.

Then, in L3, Hamlet switches topics and suggests that she doubt that truth is a liar. There is a shift from matters celestial to a banality on verity and falsity, but now, at least, all three lines share the common feature that Ophelia should doubt erroneous concepts. She should doubt that stars are Fire, that the Sun moves around the Earth and that truth is a liar. However, lines L1,2 are still orthogonal to L3 in the sense that L1,2 deal with a fairly advanced and esoteric subject of cosmology whereas L3 is a trite and elementary observation on the nature of truth. If all three lines, L1-3, advocate that Ophelia doubt erroneous matters, might not the matters in these three lines share some more fundamental property?

Another Alternative.

What if the singular "truth" in L3 refers not to the abstract notion of the general state of being-the-case but to a particular truth, just as the singular "liar" might refer to a particular prevaricator? Since the verse does not actually name the absent noun to which the attributes of "liar" pertain, playgoers are free to explore possibilities. Liars abound, but the Bard would surely provide a clue to the one he has in mind. A good guess is that L1,2 serve to guide the interpretation of the L3, in

213

which case the question becomes whether there is anything in the New Astronomy of L1,2 that is associated specifically with the liar of L3. The answer is yes, there is.

Logic.

L1,2 refer to the Old Astronomy whose greatest proponent was Aristotle, so perhaps "liar" in L3 refers to *Ille Philosophus* himself. *A Perfit Description* is one of the Bard's chief sources for *Hamlet* and the first complete statement of the new cosmic World order, wherein Digges assails Aristotle's lack of consideration of Pythagorean ideas and the resulting air of infallibility that developed about his works. By contrast, Digges' conclusions are evidence-based and, as a result, Shakespeare parodies the old ways of understanding nature.

The apparent truism, L3, contains the words "doubt," "truth," and (by implication) "falsity," which are terms that belong to the language of logical argument. In fact, the age-old matter of truth versus falsity was a major concern to Aristotle whose collected works on logic, *Organon*, constitute the grounds of the Old Philosophy. The pious philosopher's chief instrument of logic was the syllogism, and by ridiculing its axiomatic nature, Hamlet disparages the basis of his physical cosmology. He based much of his physical theory on his writings on deductive reasoning through use of syllogisms, but these are only as useful as the validity of their premises. In the sixteenth century, Petrus Ramus had tried to develop a theory of logic to supplant Aristotle's but, after fleeing France to escape religious persecution, he made the fatal error of returning to Paris only to fall victim to the St. Bartholomew's Day massacres. Reform was needed because, when Aristotle constructed his World view, he ignored the possibility that the very essence of what we today term the Copernican and Diggesian transformations, might exist. By neglecting them, he imputes falsity to them and his followers dutifully regarded the missing possibilities as unworthy of consideration.

Aristotelian logic has a serious limitation of allowing only two truth-values: perfect truth symbolized by the number 1, and absolute falsehood symbolized by the number 0. A third and embarrassingly large category contains all the statements that classical logic cannot handle at all, these being ones that are neither completely true nor completely false; i.e., have truth values between 0 and 1. A syllogistic conclusion is worth just what the premises are worth and cannot manufacture truth out of thin air. Doubt drives the cycles of scientific inquiry and, in the vast majority of instances, there is rarely enough information to lead to a single, definite, and certain conclusion. Aristotle recognized that there are shades of truth in strictly terrestrial affairs like politics and ethics, but in matters celestial, he disallows uncertainty, forgetting that fallible humans created these ideas in the first place. To this extent, the suggestion discussed above applies, that "doubt" in L3 could have the "weakened sense" of "suspect," because, strictly speaking, Hamlet himself must take care to allow the possibility of doubt in what he says, too.

In the love letter, Shakespeare uses "doubt" repetitively, instead of words more appropriate to the binary extremes of 0 and 1, such as "deny" or "accept," to demonstrate awareness of the complexity of problems in cosmology and epistemology and, in general, to show the open-endedness of inquiry in natural philosophy. Shakespeare takes aim at the philosophical implications of simplistic binary decision trees even as he takes to task the naive division of creation into natural and supernatural space. In L3, therefore, Hamlet instructs Ophelia to go against Aristotelian belief and doubt that Hamlet's methodology and transformation are false. In short, she is encouraged to accept Aristotle as a falsifier for his failures in methodology and for his neglect even of the possibility of unbounded heliocentricism. The fact that Hamlet chooses to write L3 in such a complicated way suggests that the line is a parody on devout schoolmen like Bernard of Verdun and his contemporaries in Paris whose methodology was so deficient that they brooked no alternative to Ptolemaic astronomy.

L1-3 are now closely allied and parallel in meaning since Hamlet advises Ophelia to doubt all aspects of the Old Philosophy. Whereupon, L4, being independent of the dry subjects of physical cosmology and logic, and with the sense above of "but" meaning "moreover" or "and," establishes the independent fact that Hamlet has a special literal and figurative interest in Ophelia.

Sums.

Hamlet goes on to tell Ophelia that he is "ill at these numbers" and that he has not the "art to reckon his groans." Shakespeare implies that the word "ill" means "artless," which agrees with the meanings "unskillful" and "inexpert" in *OED*. In a standard reading, therefore, "numbers" could refer to the artless words and lines of the letter. Alternatively, the word "stars" occurring in L1 is the only plural noun antecedent to "these numbers." L1 refers to the "fire" of the stars, but the love letter does not mention the distribution or number of stars. Someone, like Digges, with access to a telescope, might well groan at the sheer numbers of stars that he can see and whose incidence he, as a thorough investigator of nature, might feel obliged to count. Therefore, "I am ill at these numbers" joins L2 in accounting for both essential aspects of the New Astronomy, viz. that the Sun lies motionless at the center of the planetary system and that innumerable Fixed Stars populate infinite space.

Frame.

Hamlet signs off his letter by calling himself "this machine." Critics agree that "machine" refers to Hamlet's body, which is a sense of the word unique in the Canon. The *OED* cites the example of a human or animal frame seen "as a combination of several parts" each performing a function. The young man sees himself as fully functional, a fact of interest to Ophelia no doubt, but the general sense is that he pledges devotion while his "bodily frame belongs to him" (*OED* 4c; Jenkins 2.2.122-3n). Hamlet reduces love to mechanical terms and comes off looking like an unfeeling and artless suitor, although, in principle, "this machine" could refer also to Ophelia's frame, in which case Hamlet's ambiguous pronouncement of devotion "whilst this machine is to him" means that he pledges devotion to her for as long as she "is (devoted) to him." It is no surprise that, in 3.1, as soon as Hamlet suspects her of dishonesty he breaks off their relationship and suggests she live a cloistered life.

"Machine" is also "a structure of any kind, material or immaterial, a fabric, an erection." The *OED* cites an example from the 1599 hymn *Of Gods Benefites Bestowed vpon Man* by the Scottish poet and churchman

Alexander Hume (1557?-1609). In the passage below, the first two lines, 37-8, deal with the creation of the Universe as related in *John* 1.1 (Lawson 18). Hume's lengthier account of the creation according to *Genesis* 1 begins in the next two lines 39-40:

> Euen be his wisedome, and his word, sa wondrouslie of nocht,
> This machine round, this vniuers, this vther warld he wrocht.
> He creat first the heauen, the earth, and all that is thairin,
> The swelling seas, the fire, and aire, sine man deuoid of sinne.

The description "machine" is apt since the Ptolemaic model, with all its rods and wheels, resembles an intricate mechanical device. Thus, following Hume, Hamlet's "machine" could be the frame or structure of the Universe. This is the sense used by Digges a quarter-century earlier, in which in this form or frame of the Universe we behold a wonderful *Symmetry* of motions. In scene 2.2 in which Polonius reads Hamlet's letter, Hamlet refers to cosmic existents using architectural terminology, "this goodly frame, the earth," "this canopy the air," "this brave o'erhanging firmament," and "this majestical roof fretted with golden fire." Hamlet does not stand on romantic niceties because the cold, hard facts of natural philosophy per se have nothing to do with emotion and do not respect feelings. In the abstract world of allegory, alignments of the heart play second fiddle to alignments of heavenly bodies and the music of the spheres.

Therefore, when Hamlet writes of the new model being "to him" he is speaking allegorically. When he signs off, he announces, "there is a great deal to him" – by which, he means the New Philosophy! The association of "machine" with Hamlet is not hard to accept because, by hypothesis, the two components of the New Philosophy are an integral part of his dramatic construction. Altogether, Hamlet's closing sentence means that he will remain faithful to Ophelia for as long as she remains true to him and as long as he is alive to personify the two chief components of the New Astronomy.

The irony is that, in reading Hamlet's letter to the royal couple, Polonius unknowingly presents a World view that is antithetical to the one that his geocentricist monarch holds. Irony has victims and Polonius and the king are they. Irony brings opposing theories into direct conflict, and here Shakespeare succeeds in making the bigoted advocates of bounded geocentricism look foolish.

Fabrication.

The interpretations above face the difficulty that Ophelia had told her father previously that she refused to accept anything from Hamlet and, in particular, that she refused his letters. Probably, the ingénue is not lying. As noted above, perhaps she already had the love letter when she made this promise, but there is no textual basis for the suggestion.

In an unpublished paper discovered after his death, Goddard (*College*) expresses suspicion of the love letter's source, noting that the quality of the writing is not what one expects from a sophisticated Prince and that Polonius skillfully dodges the question of authorship. Goddard regards Polonius' presentation as excessively devious and feels that L3 "points straight at Polonius" because he is the sort of person who blurs the truth. Goddard argues that Act 2 opens with over 70 lines in which Polonius instructs Reynaldo on how to spy on his son, and that Polonius has a "despotic mind in its most cowardly aspect." Polonius gives Reynaldo permission to burden Laertes with "what forgeries" he pleases, i.e., to inflict "invented matters" upon him (Edwards 2.1.20n), and the only other use of "forgery" in *Hamlet* is spoken in 4.7 by the spymaster's deceitful partner, Claudius. Polonius' suspicions of Laertes run true to form because they spring from what he thinks, or fears, is true, and not from empirical reality. Like Rosencrantz and Guildenstern, Polonius is an expert at leaping to false conclusions, and Goddard concludes that the Reynaldo scene "clinches the idea that Polonius was capable of forgery." At least one other critic agrees, terming Goddard's theory "ingenious" (Erlich 220).

Polonius has the motive to forge a love letter because he wishes to prevent Hamlet from seeing his daughter, but to avoid confrontation he manipulates Hamlet's next-of-kin into doing the job for him. The trouble is that Polonius has only the say-so of his daughter to back him up. Mere absence of hard evidence is not a problem to the scurrilous spymaster who simply manufactures what he needs. The idea of a letter probably came to him after Ophelia told him that she did not accept any letters from Hamlet. Knowing of Hamlet's amorous interests and his attempt to write to Ophelia, Polonius had the bright idea of composing a letter, ostensibly from Hamlet, which would confirm his diagnosis of madness. Polonius has time to devise the plan in the interval between scenes 2.1 and 2.2 and during the first 39 lines of 2.2 when he is not on stage. It is essential, then, that Polonius not bring Ophelia with him when he reports to the king lest she inadvertently blurt out the truth.

Gifts.

When Act 3 begins, sufficient time has elapsed for father and daughter to get their story straight, and events in the nunnery scene, 3.1, support the theory. While the two "lawful espials," Polonius and the king, eavesdrop, Hamlet enters and launches into an existential monologue, following which Ophelia tries to get Hamlet to take back "remembrances" that, she says, he gave to her. He denies giving her any, but she insists that he did and says, moreover, that with his gifts came "words of so sweet breath composed / As made the things more rich." It is unclear whether Ophelia's "words" refer to ones spoken or written (Edwards 3.1.98n). If written, then perhaps he did in fact give her the letter that Polonius bears. If spoken, then it looks as if Ophelia is putting us on.

Despite Hamlet's denials, Ophelia unloads the alleged love-tokens with the words, "There my Lord." One envisions her plunking down the artifacts on some handy surface, making it sound as if she is glad to be rid of them. The only time Ophelia showed any real spunk was early on, in 1.3, when she attempted to defend Hamlet's interest in her and before her father circumscribed her life and quashed her free will. Ophelia's sudden show of assertiveness is out of character, suggesting that Polonius has emboldened her. She follows his instructions well enough because, after all, she cannot be blamed if Hamlet does not actually pick up the artifacts. With Ophelia doing some acting of her own and her father listening in, it is likely the two conspire to support the idea that the love letter is real.

Forgery.

Some portray Polonius as wise and dignified, but he lacks conscience and empathy. He has poor judgment and, like the geocentric monarch he serves, is pathologically egocentric. The chamberlain wishes that his daughter and her suitor would play the roles he expects of them, just like the Sun and Moon that dutifully hold their courses and do what they're supposed to do, which is to light the center of his Universe, the Earth. Polonius is completely out of touch with cosmic reality and, in writing the letter, he knows no better than to try to compromise Hamlet's sanity by having him advise Ophelia to deny the standard cosmology of the day. The vulgarity and zealousness of the writing uncovers the way Polonius thinks. He may be an inherently inept writer, but perhaps he deliberately affects a disjointed, puerile style, the better to paint Hamlet as the court lunatic.

To Polonius, perfection resides in the heavens, so he has Hamlet call Ophelia "celestial" in order to show how much the Prince adores her. Polonius writes from the heart of his own belief-structure again when, in L1,2, he makes it sound as if Hamlet is denying the truth of the standard World view. The third line L3 contains a platitude so pedantic as to impugn Hamlet's sanity. In L4, with the sense of "but" meaning "also," Polonius delivers the punch line that Hamlet loves Ophelia. Lines L1-3 establish the illiterate quality of the letter and Hamlet's insanity, and L4 plus Hamlet's mechanical virtues establish the base quality of Hamlet's love. It appears that Polonius has succeeded in defining Hamlet as madly in love.

Goddard's hypothesis leads to the interpretation that, by forging a letter purportedly from Hamlet to Ophelia, Polonius unknowingly advances the New Philosophy under the very noses of the staunch defenders of the standard World view. Aristotle believed that the gods could see everything, a trait that allowed them to serve dramatic functions, the irony here being that Shakespeare lets supernatural powers have a little fun at the expense of pedantry through the dramatic convenience of guiding Polonius' hand as he pens the letter. Moreover, in indicting Aristotelianism, Shakespeare uses irony ironically since, in *Poetics,* Aristotle is himself an advocate of irony.

AFTERWORD

Chapters 6-12 make the case that *Hamlet* has a sub-text documenting the state of theoretical and empirical cosmology extant at the turn of the seventeenth century. By addressing evidence-based inquiry, Shakespeare turns *Hamlet* into a celebration of the advent of the New Philosophy. In addition, the reverence implicit in *Hamlet* brings into focus the puzzle concerning the nature of physical and supernatural space, and is but one instance of the widespread perception that the human mind instinctively places divine power in heaven. Literature, prayer, and song consistently attest to the belief that humans yearn for unity with the Progenitor of creation. Plato does so in *Laws* and *Timaeus*, as does Aristotle at the end of *De Caelo*. In *Critique of Pure Reason*, Immanuel Kant (1724-1804) remarks, in a celebrated passage, "Two things fill the mind with ever new and increasing wonder and awe ... the starry heavens above me and the moral law within me."

Faith and reason are closely associated in both *Hamlet* and *A Perfit Description*. In 4.4, Shakespeare speaks of the Creator's gift to humankind of god-like reason, which is that special human attribute that renders humans capable of desiring union with the Almighty. Digges believes that humankind is god-like by virtue of intelligence, "Why shall we so much dote in the appearance of our sences, which many ways may be abused, and not suffer our selues to be directed by the rule of Reason, which the great GOD hath giuen us as a Lampe to lighten the darkness of our understanding." To Thomas Digges, understanding natural existents is impossible without simultaneous consideration of supernatural existence. He makes the point by capitalizing "GOD" in the quote above. Similarly, in *Merchant of Venice*, Old Gobbo invokes the deity, and the compositors of the first and second printings capitalized "GOD" there as well, "as if in a devotional work" (Brown pp. xi, xviii). Galileo echoes the sentiment, asserting that God endowed humans with senses, reason and intellect by which to gain knowledge of the World.

Digges and Shakespeare address cosmology both scientifically and theologically, which is not to say that they embrace argument from design. Shakespeare does not appeal to divine action to "save" physical phenomena for which no scientific explanations exist. Rather, he leaves the task of understanding the Universe to future generations of rational thinkers. He expresses confidence that the quest will succeed by using the word "infinite" to describe both cosmic space and the human faculty whose challenge is to understand it. As Hamlet puts it, "What a piece of work is a man! How noble in reason, how infinite in faculties, in form and moving how express and admirable, in action how like an angel, in apprehension how like a god!" The message is one of hope that, someday, humankind will understand creation and being-in-the-world.

WORKS CONSULTED

Allen, Richard Hinckley. *Star Names: Their Lore and Meaning* (1899). New York: Dover, 1963.

Altena, van Regteren I.Q. *Jacques de Gheyn: Three Generations*. 3 Vols. The Hague: Nijhoff, 1983.

Altman, Linda Jacobs. *A History of Infectious Disease*. Berkeley Heights, NJ: Enslow, 1998.

Altrocchi, Paul H. "Sleuthing an enigmatic Latin annotation." *Shakespeare Matters* 2, No. 4 (2003), 16-9.

Altschuler, Eric, and William Jansen. News Notes. *The Oxfordian* 8 (2005), 153-4.

American Astronomical Society Newsletter Number 84, March (1997), 16.

Andrews, John F., ed. *Hamlet*. By William Shakespeare. London: Dent, 1993.

Asimov, Isaac. *Guide to Shakespeare* (1970). 2 vols. New York: Random House, 1993.

Aughterson, Kate, ed. The *English Renaissance: An Anthology of Sources and Documents*. New York: Routledge, 1998.

Baade, Walter. "B Cassiopeiae as a Supernova of Type I." *The Astrophysical Journal* 102 (1945), 309-317.

Bacon, Roger. *Fryer Bacon his discovery of the miracles of art, nature, and magick faithfully translated out of Dr. Dees own copy by T.M. and never before in English* (*c.*1250). London: Simon Miller, 1659.

Bakeless, John E. *Christopher Marlowe*. New York: Morrow, 1937.

Baldwin, T. W. *William Shakspere's Small Latine & Lesse Greek*. 2 vols. Urbana: University of Illinois Press, 1944.

Barnes, Larry. "Canker." *Encyclopedia of Plant Pathology*. New York: Wiley, 2001, pp. 175-6.

Barnet, Sylvan. "Shakespeare: Prefatory Remarks." In *The History of Troilus and Cressida*, by William Shakespeare. Ed. Daniel Seltzer. New York: Signet, 1988, pp. vii-xxi.

Beauregard, David N. "Shakespeare against the Skeptics: Nature and Grace in The Winter's Tale." *Shakespeare's Last Plays: Essays in Literature and Politics*. Eds. Stephen W. Smith and Travis Curtright. Lanham MD: Lexington Books, 2002, pp. 53-72.

Beaurline, Lester A., ed. *King John*. By William Shakespeare. Cambridge: Cambridge University Press, 1990.

Beckingsale, B.W. *Burghley: Tudor Statesman 1520-1598*. London: Macmillan, 1967.

Bellow, Saul. "Deep Readers of the World, Beware." *Herzog*. New York: Viking, 1976.

Berry, Arthur. *A Short History of Astronomy* (1898). New York: Dover 1961.

223

Best, Jason S., Sara A. Maene, and Peter Usher. "Copernicus's Neglected Successor." *Mercury Magazine* 30, No. 5 (2001), 38-40.

Bevington, David, ed. *The Complete Works of Shakespeare*. New York: Harper Collins, 1992.

Birringer, Johannes H. *Marlowe's Dr. Faustus and Tamburlaine: Theological and Theatrical Perspectives*. Frankfurt-am-Main: Peter Lang, 1984.

Blessing, Lee. "Patient A." In *Four Plays*. Portsmouth, NH: Heinemann, 1995.

Blundeville, Thomas. *Theoriques of the Seuen Planets ...* London: Adam Islip, 1602.

Boorstin, Daniel J. *The Discoverers*. New York: Random House, 1983.

Brandt, John C. and Stephen P. Maran. *New Horizons in Astronomy*. San Francisco: Freeman, 1979.

Braunmuller, A.R. ed. *The Life and Death of King John*. By William Shakespeare. Oxford: Oxford University Press, 1994

Breight, Curtis C. *Surveillance, Militarism and Drama in the Elizabethan Era*. New York: St. Martin's Press, 1996.

Brennecke, Ernest. *Shakespeare in Germany 1590-1700*. Chicago: University of Chicago Press, 1964.

Brewer, D.S., ed. *Geoffrey Chaucer: The Parlement of Foulys*. London: Nelson, 1960.

Brown, John Russell, ed. *The Merchant of Venice*. By William Shakespeare. London: The Arden Shakespeare, 2004.

Budiansky, Stephen. *Her Majesty's Spymaster: Elizabeth I, Sir Francis Walsingham, and the Birth of Modern Espionage*. London: Viking Penguin, 2005.

Bulfinch, Thomas. *Mythology*. New York: Avenel, 1978.

Bullough, Geoffrey. *Narrative and Dramatic Sources of Shakespeare*. New York: Columbia University Press, 1973.

Burke, Robert Belle, trans. *The Opus Majus of Roger Bacon*. New York: Russell and Russell, 1962.

Burton, Robert. *The Anatomy of Melancholy* (1620). New York: Tudor, 1951.

Cahn, Victor L. *Shakespeare the Playwright*. New York: Greenwood Press, 1991.

Campbell, Lily B. *Shakespeare's "Histories" Mirrors of Elizabethan Policy*. San Marino: The Huntington Library, 1947.

Carefoot, G. L., and E. R. Sprott. *Famine on the Wind: Man's Battle Against Plant Disease*. Chicago: Rand McNally, 1967.

Carey, John. "Introduction." In *John Donne*. Ed. John Carey. Oxford: Oxford University Press, 1990.

Charney, Maurice. *Hamlet's Fictions*. New York: Routledge, 1988.

Chartrand, Mark R. III. *Skyguide: A Field Guide to the Heavens*. New York: Golden Press, 1982.

Christianson, John R. *On Tycho's Island: Tycho Brahe and his Assistants*. Cambridge: Cambridge University Press, 2000.

Clark, Cumberland. *Shakespeare and Science*. Birmingham: Cornish Brothers, 1929.

Clark, Peter. *English Provincial Society from the Reformation to the Revolution: Religion, Politics, and Society in Kent 1500-1640*. Rutherford: Fairleigh Dickinson Press, 1977.

Clarke, Charles and Mary Cowden. *The Shakespeare Key: Unlocking the Treasures of his Style* ... London: Samson Low et al., 1879.

Clegg, Brian. *The First Scientist: A Life of Bacon*. London: Constable, 2003.

Coffin, Charles M. *The Complete Poetry and Selected Prose of John Donne*. New York: The Modern Library, 1952.

- - - . *John Donne and the New Philosophy*. New York: Columbia University Press, 1937.

Connolly, John M., and Thomas Keutner. "Introduction: Interpretation, Decidability, Meaning." In *Hermeneutics vs. Science: Three German Views*. J.M. Connolly and T. Keutner, eds. Notre Dame: University of Notre Dame Press, 1988, pp. 1-67.

Copernicus, Nicolaus. *On the Revolutions of the Heavenly Spheres* (1543). Trans. Charles Glenn Wallis. New York: Prometheus, 1995.

Crewe, Jonathan, ed. *Troilus and Cressida*. By William Shakespeare. New York: Penguin, 2000.

Crombie, A.C. *The History of Science from Augustine to Galileo* (1952). 2 vols. New York: Dover, 1995.

Crompton, Samuel. "The Portrait of Tycho Brahe." *Nature* XVI (1877), 501-2.

Crowe, Michael J. *Theories of the World from Antiquity to the Copernican Revolution*. 2nd edn. Mineola, NY: Dover, 2001.

Crystal, David, and Ben Crystal. *Shakespeare's Words: A Glossary and Language Companion*. London: Penguin, 2002.

Dampier, William C. *A History of Science and its Relation with Philosophy and Religion* (1929). Cambridge: Cambridge University Press, 1961.

Darius, Jon. "Report of Discussion." *Bulletin of the Scientific Instrument Society* No. 37 (1993), 6, 10.

Davis, Frank M. "Shakespeare's Medical Knowledge: How did he acquire it?" *The Oxfordian* 3 (2000), 45-58.

Deacon, Richard. *John Dee*. London: Frederick Muller, 1968.

Dent, Alan. *The World of Shakespeare*. New York: Taplinger, 1979.

Dessen, Alan C. "Weighing the Options in Hamlet Q1." In *The Hamlet First Published (Q1, 1603)*. Ed. Thomas Clayton. Newark, University of Delaware Press, 1992, pp. 65-78.

Dicks, David R. *Early Greek Astronomy to Aristotle*. New York: Cornell University Press, 1970.

Diehl, Huston. *An Index of Icons in English Emblem Books 1500-1700*. Norman: University of Oklahoma Press, 1986.

Digges, Leonard. *A Prognostication Everlasting Corrected and Augmented by Thomas Digges* (1576). Amsterdam: Theatrum Orbis Terrarum, 1975.

Digges, Thomas. *Stratioticos* (1579). New York: Da Capo Press, 1968.

Dijksterhuis, E.J. *The Mechanization of the World Picture* (1961). Princeton: Princeton University Press, 1986.

Durant, Will. *The Story of Philosophy*. New York: Simon and Schuster, 1961.

Drake, Stillman, trans. *Discoveries and Opinions of Galileo*. New York: Doubleday, 1957.

- - - . *Galileo at Work: His Scientific Biography* (1978). New York: Dover, 1995.
- - - . "Galileo: A Biographical Sketch." In *Galileo: Man of Science*. Ed. Ernan McMullen. Princeton: The Scholar's Bookshelf, 1988.
- - - . *Telescopes, Tides, and Tactics*. Chicago: University of Chicago Press, 1983.
Dreyer, John L.E. *A History of Astronomy from Thales to Kepler* (1906). New York: Dover, 1953.
- - - . *Tycho Brahe* (1890). Gloucester (Mass.): Peter Smith, 1977.
- - - , ed. *Tychonis Brahe Dani Opera Omnia*. 15 vols. Hauniae Libraria Gyldendaliana, 1913-29.
Durham, Frank, and Robert D. Purrington. *Frame of the Universe: A History of Physical Cosmology*. New York: Columbia University Press, 1983.
Easton, Joy B. "Digges, Thomas." In *Dictionary of Scientific Bibliography*. Ed. Charles Coulston Gillespie. New York: Scribner's, 1971.
Edelman, Charles. *Shakespeare's Military Language*. London: Athlone, 2000.
Edmond, Mary. *Hilliard and Oliver*. London: Robert Hale, 1983.
Edwards, Philip, ed. *Hamlet, Prince of Denmark*. By William Shakespeare. Cambridge: Cambridge University Press, 1995, 2003.
Elton, Oliver, trans. *The First Nine Books of the Danish History of Saxo Grammaticus*. London: Nutt, 1894.
Erlich, Avi. *Hamlet's Absent Father*. Princeton: Princeton University Press, 1977.
Everitt, Alan. *Continuity and Civilization: The Evolution of Kentish Settlement*. Leicester: Leicester University Press, 1986.
Farley-Hills, David. "Introduction." In *Critical Responses to Hamlet 1600-1790*. Ed. David Farley-Hills. AMS Press, 1994.
Fisher, Peter, trans. *Saxo Grammaticus and the History of the Danes*. Ed. Hilda Ellis Davidson. Cambridge: Brewer, 1979.
Fix, John D. *Astronomy* 2nd edn. Boston: McGraw-Hill, 1999.
Flannagan, Roy. *John Milton: Paradise Lost*. New York: Macmillan, 1993.
Fleischer, Martha Hester. *The Iconography of the English History Play*. Salzburg: Institut für Englische Sprache und Literatur, 1974.
Fleissner, Robert F. *The Shakespeare Newsletter* 53, No. 3 (2003), 73, 96, 98.
Fletcher, Angus. "Allegory in Literary History." In *Dictionary of the History of Ideas*. New York: Scribner's, 1973. Vol. 1, pp. 41-8.
Forker, Charles R. ed. *King Richard II*. By William Shakespeare. London: The Arden Shakespeare, 2004.
Franklin, Benjamin. *Poor Richard's Almanac*. New York: Dodge Publishing, 1757.
Freire, Paulo. *Pedagogy of the Oppressed*. New York: Continuum, 1993.
Frisch, Harold. *Hamlet and the Word: The Covenant Pattern in Shakespeare*. New York: Ungar, 1971.
Furness, Horace Howard, ed. *A New Variorum Edition of Shakespeare*. Philadelphia: Lippencott, 1877. Vol. 3.
Gade, John Allyne. *The Life and Times of Tycho Brahe*. Princeton: Princeton University Press, 1947.
Garber, Marjorie. *Shakespeare After All*. New York: Anchor, 2004.

Gardner, Helen Louise, ed. *The Metaphysical Poets*. London: Penguin, 1985.

Garrod, Heathcote William, ed. *John Donne: Poetry and Prose*. Oxford: Clarendon Press, 1946.

Gatti, Hilary. *The Renaissance Drama of Knowledge*. London: Routledge, 1989.

Gettings, Fred. *The Secret Zodiak: The Hidden Art in Mediaeval Astrology*. London: Routledge and Kegan Paul, 1987.

Ginsburg, I. "The disregard syndrome: A menace to honest science?" *The Scientist* 15, No. 24, (2001), 51.

Gingerich, Owen, and Robert S. Westman. "The Wittich Connection: Conflict and Priority in Late Sixteenth-Century Cosmology." *Transactions of the American Philosophical Society* 78, Part 7 (1988).

Goddard, Harold C. *The Meaning of Shakespeare*. 2 vols. Chicago: University of Chicago Press, 1951.

- - - . "Hamlet to Ophelia." *College English*, 16, 403-15 (1955).

Gollancz, Israel. *The Sources of Hamlet: With Essay on the Legend*. London: H. Milford, Oxford University Press, 1926.

Gorton, Lisa. "John Donne's Use of Space." *Early Modern Literary Studies* 4.2 (1998).

Gottschalk, Paul. *The Meanings of Hamlet*. Albuquerque: University of New Mexico Press, 1972.

Gould, Laurence M. "Introduction." In *George Gaylord Simpson. Attending Marvels: A Patagonian Journal*. New York: Time, 1965, pp. xvii-xxi.

Grant, Edward. *Physical Sciences in the Middle Ages*. Cambridge: Cambridge University Press, 1977.

Graves, Robert. *The Greek Myths*. 2 vols. London: Penguin Books, 1960.

Green, David A., and F. Richard Stephenson. "The Historical Supernovae." In Supernovae and Gamma Ray Bursters. Ed. Kurt W. Weiler. Lecture Notes in Physics. Berlin: Springer-Verlag, 2003. Vol. 598, pp. 7-19.

Guilfoyle, Cherrell. *Shakespeare's Play within Play*. Kalamazoo: Western Michigan University Medieval Institute, 1990.

Gurr, Andrew. *Hamlet and the Distracted Globe*. Edinburgh: Scottish Academic Press and Sussex University Press, 1978.

Hadsund, Per. "The tin-mercury mirror: its manufacturing technique and deterioration process." *Studies in Conservation* 38 (1993), 3-16.

Hakluyt, Richard. *Voyages in Search of the North-West Passage*. Ed. Henry Morley. New York, Cassell, 1886.

Hall, Marie Boas. "Science." In *The New Cambridge Modern History* (1957-79). Vol. 3 ed. Richard B. Wernham. Cambridge: Cambridge University Press, 1968, pp. 453-79.

Halliwell, James Orchard, ed. *A Collection of Letters Illustrative of the Progress of Science in England from the Reign of Queen Elizabeth to that of Charles the Second*. London: R. and J.E. Taylor, 1841.

Halstead, Ron. *Shakespeare Matters* 4, No. 2 (2005), 10.

Hamill, John. "The Ten Restless Ghosts of Mantua." *Shakespeare Oxford Newsletter* 39 (2003), 3-6 and 18-9.

Hankins, John E. *Shakespeare's Derived Imagery*. Lawrence: University of Kansas Press, 1967.

Hansen, William F. *Saxo Grammaticus & the Life of Hamlet*. Lincoln: University of Nebraska Press, 1983.

Hardy, Evelyn. *Donne, A Spirit in Conflict*. London: Constable, 1942.

Hariot, Thomas. *A briefe and true report of the new found land of Virginia . . .* (1588). See Quinn 317-87.

Harrison, Edward R. *Cosmology: The Science of the Universe*. Cambridge: Cambridge University Press, 1981.

Healy, Paul J. "The Copernican Dimension in the Kantian Revolution." *The Astronomy Quarterly* 5 (1987), 197-206.

Heath, Thomas L. *Aristarchus of Samos: The Ancient Copernicus* (1913). New York: Dover 1981.

- - - . *Greek Astronomy* (1932). New York: Dover, 1991.

Helm, P.J. *England under the Yorkists and Tudors 1471-1603*. New York: Humanities, 1968.

Heninger, S.K. Jr. *Touches of Sweet Harmony: Pythagorean Cosmology and Renaissance Poets*. San Marino: Huntington Library, 1979.

Hernadi, Paul. *Beyond Genre: New Directions in Literary Classification*. Ithaca: Cornell University Press, 1973.

Hetherington, Barry. A *Chronicle of Pre-Telescopic Astronomy*. Chichester: Wiley, 1996.

Hetherington, Noriss S. "Ptolemy: on trial for fraud." *Astronomy and Geophysics* 38, No. 2 (1997), 24-7.

Hibbard, George R., ed. *Hamlet*. By William Shakespeare. New York: Oxford University Press, 1987.

- - - . *Thomas Nashe*. London: Routledge and Paul, 1962.

Hockney, David. *Secret Knowledge*. London: Thames and Hudson, 2001.

Hoff, Linda Kay. *Hamlet's Choice*. Lewiston: Edward Mellen, 1988.

Holland, Clive. *Denmark: the land of the sea kings*. London: A. and C. Black, 1928.

Honigmann, Ernst A.J., ed. *King John*. By William Shakespeare. London: Routledge, 1994.

Hoskin, Michael A. *The Cambridge Illustrated History of Astronomy*. Cambridge: Cambridge University Press, 1997.

- - - . *The History of Astronomy*. Oxford: Oxford University Press, 2003.

Hotson, Leslie. *I, William Shakespeare Do Appoint Thomas Russell, Esquire ...* New York: Oxford University Press, 1938.

- - - . *Shakespeare by Hilliard*. Berkeley: University of California Press, 1977.

Howse, Derek. Letter. *Bulletin of the Scientific Instrument Society* No. 37 (1993), 9.

Hughes, Merritt Y. "Kidnapping Donne." In *Essential Articles for the study of John Donne's Poetry*. Ed. John R. Roberts. Hamden: Archon Books, 1975, pp. 37-57.

Hughes, Stephanie H. "Shakespeare's Tutor: Sir Thomas Smith." *The Oxfordian* 3 (2000), 19-44.

- - - . Editor's Notes. *The Oxfordian* 8 (2005), 107n2 and 108n8.

Huizinga, Johan. *Tien Studiën*. Haarlem: Tjeenk Willink, 1926.

Hunter, G.K. ed. *All's Well That Ends Well*. By William Shakespeare. London: The Arden Shakespeare, 2000.

Jaki, Stanley L. *The Milky Way*. New York: Science History Publications, 1972.

Janowitz, Henry. "Some Evidence on Shakespeare's Knowledge of the Copernican Revolution and the 'New Philosophy'." *The Shakespeare Newsletter* Vol. 51 No. 3 (2001), 79-80.

Jenkins, Harold, ed. *Hamlet*. By William Shakespeare. Nelson: Walton-on-Thames, 1997.

Johnson, Francis R. *Astronomical Thought in Renaissance England: A Study of English Scientific Writings from 1500 to 1645*. Baltimore: Johns Hopkins, 1937.

- - - . "Latin versus English: The Sixteenth-Century Debate over Scientific Terminology." *Studies in Philology* Vol. XLI, No.2 (1944), 109-35.

Johnson, Francis R., and Sanford V. Larkey. "Thomas Digges, the Copernican System, and the Idea of the Infinity of the Universe in 1576." *The Huntington Library Bulletin* No. V (1934), 69-117.

Johnston, William Preston. *The Prototype of Hamlet*. New York: Belford, 1890.

Jones, Harold Spencer. *General Astronomy*. London: Arnold, 1951.

Kelly, Philippa. "Surpassing Glass: Shakespeare's Mirrors." *Early Modern Literary Studies* 8.1 (2002).

Kippis, Andrew, ed. *Biographia Britannica*. 5 vols. London: Strahan, 1793.

Kitto, H.D.F. *Form and Meaning in Drama*. London: Methuen, 1956.

Knobel, E.B. "Astronomy and Astrology." In *Shakespeare's England: An Account of the Life and Manners of his Age*. Eds. Walter Alexander Rayleigh, Sidney Lee, and Charles Talbut Onions. Oxford: Clarendon, 1950, 444-61.

Knowles, David. "Bernard of Clairvaux, St." In *The Encyclopedia of Philosophy*. Ed. Paul Edwards. New York: Macmillan, 1967.

Kocher, Paul H. *Christopher Marlowe: A Study of his Thought. Learning, and Character*. New York: Russell and Russell, 1962.

Kosso, Peter. *Reading the Book of Nature*. Cambridge: Cambridge University Press, 1992.

Koyré, Alexandre. *The Astronomical Revolution: Copernicus - Kepler - Borelli*. Trans. R.E.W. Maddison. Ithaca: Cornell University Press, 1973.

- - - . *From the Closed World to the Infinite Universe*. Baltimore: Johns Hopkins University Press, 1957.

Kuhn, Thomas S. *The Copernican Revolution: Planetary Astronomy in the Development of Western Thought*. Cambridge: Harvard University Press, 1957.

- - - . *The Structure of Scientific Revolutions*. Chicago: University of Chicago Press, 1970.

Lawson, Alexander. *The Poems of Alexander Hume*. Edinburgh: Blackwood, 1902.

Leonard, John, ed. *John Milton Paradise Lost*. London: Penguin, 2000.

Levi, Peter. *The Life and Times of William Shakespeare*. New York: Holt, 1988.

Levin, Richard. "Frailty, Thy Name is Wanton Widow." *The Shakespeare Newsletter* 55, No.1 (2005), 5-6.

Levy, David H. *The Man who Sold the Milky Way: A Biography of Bart Bok*. Tucson: University of Arizona Press, 1993.

- - - . "Shakespeare's Eclipses Return." *Sky and Telescope* 103, No. 6 (2002), 75.

- - - . *Starry Night: Astronomy and Poets Read the Sky*. Amherst, NY: Prometheus, 2001.

Lewis, C.S. *The Discarded Image*. Cambridge: Cambridge University Press, 1964.

Lindberg, David C. *Theories of Vision from Al-Kindi to Kepler*. Chicago: University of Chicago Press, 1976.

Loades, D.M. *Two Tudor Conspiracies*. Cambridge: Cambridge University Press, 1965.

Luscombe, David. In *The Encyclopedia of Philosophy*. Ed. Paul Edwards. New York: Macmillan, 1967.

Maclean, Hugh, ed. *Edmund Spenser's Poetry*. New York: Norton, 1982.

Maene, Sara A., Jason S. Best, and Peter D. Usher. "The Incidence of Sixteenth Century Cosmic Models in Modern Texts." *Bulletin of the American Astronomical Society* 31 (1999), 1530.

Martz, William J., ed. *Hamlet*. By William Shakespeare. Glenview IL: Scott Foresman, 1970.

Matiegka, Heinrich. *Bericht über die Untersuching der Gebeine Tycho Brahe's*. Prague: Gesellschaft der Wissenschaften, 1901.

Matsuura, Kaichi. *A Study of Donne's Imagery*. Tokyo: Kenkyusha, 1972.

Matthews, Boris. *The Herder Symbol Dictionary*. Wilmette: Chiron, 1986.

May, Steven W. *The Elizabethan Courtier Poets*. Columbia: University of Missouri Press, 1991.

- - - . "The Poems of Edward DeVere ... and of Robert Deveraux ..." *Studies in Philology* LXXVII, No. 5, 1980.

McAlindon, T. *Shakespeare's Tragic Cosmos*. Cambridge: Cambridge University Press, 1991.

McKerrow, Ronald B. ed. *The Works of Thomas Nashe*. Oxford: Blackwell, 1958.

McLean, Antonia. *Humanism and the Rise of Science in Tudor England*. New York: Neale Watson Academic Publishers, 1972.

Meadows, Arthur J. *The High Firmament*. Leicester: Leicester University Press, 1969.

Melchiori, Giorgio. "Hamlet: The Acting Version and the Wiser Sort." In *The Hamlet First Published (Q1, 1603)*. Ed. Thomas Clayton. Newark: University of Delaware Press, 1992, pp. 195-210.

Michell, John. *Who Wrote Shakespeare?* London: Thames and Hudson, 1996.

Mills, Allan A. "Did an Englishman Invent the Telescope?" *Bulletin of the Scientific Instrument Society* No. 35 (1992), 2.

Mitton, Jacqueline. *The Penguin Dictionary of Astronomy*. London: Penguin, 1993.

Moran, Gordon. *Silencing Scientists and Scholars in Other Fields*. Greenwich, Conn: Ablex Publishing, 1998.

Mortensen, Harold. "Portraeter af Tycho Brahe." *Cassiopeia* 8 (1946), 53-77.

Mowat, Barbara, and Paul Werstine, eds. *Hamlet.* By William Shakespeare. New York: Washington Square Press, 1992.

- - - . *Love's Labour's Lost*. New York: Washington Square Press, 1996.

Mousley, Andrew. In *John Donne*. Ed. Andrew Mousley. New York: St. Martin's Press, 1999.

Newton, R.R. *The Crime of Claudius Ptolemy*. Baltimore: Johns Hopkins University Press, 1977.

Nicholl, Charles. *The Reckoning: The Murder of Christopher Marlowe*. Chicago: University of Chicago Press, 1992.

Nikiforuk, Andrew. *The Fourth Horseman of the Apocalypse*. New York: Evans, 1991.

Nockolds, Peter. *The Times* (London) 16 Jan. 1997, p. 19.

North, John David. *Chaucer's Universe*. Oxford: Clarendon, 1988.

- - - . *The Measure of the Universe*. New York: Dover, 1990.

Ogburn, Charlton. *The Mysterious William Shakespeare: The Myth and the Reality*. McLean, Virginia: EPM, 1992.

Olson, Donald W., Marilyn S. Olson, and Russell L. Doescher. "The Stars of Hamlet." *Sky and Telescope* 96, No. 5 (1998), 67-73.

Orlob, G.B. "Ancient and medieval plant pathology." *Pflanzenschutz-Nachrichten* Bayer. Leverkusen: Bayer, 1973.

Osterbrock, Donald E. "Physical State of the Emission-Line Region." *Physica Scipta*, 17 (1978), 285-92.

Otté, Elise C. *Scandinavian History*. London: Macmillan, 1874.

Pacholczyk, A.G. *The Catastrophic Universe: An Essay in the Philosophy of Cosmology*. Tucson: Pachart, 1984.

Palingenius, Marcellus. *The Zodiake of Life* (1565). Trans. Barnabe Googe. New York: Scholars' Facsimiles & Reprints, 1947.

Palmer, Alan, and Veronica Palmer. *Who's Who in Shakespeare's England*. New York: St. Martin's Press, 1981.

Palmer, Richard E. *Hermeneutics*. Evanston: Northwestern University Press, 1969.

Panek, Richard. *Seeing and Believing*. New York: Viking, 1998.

Pannekoek, Anton. *A History of Astronomy* (1961). New York: Dover, 1989.

Patterson, Louise. Letter to the Editor. *Isis* 43 (1952), 122.

Payne-Gaposchkin, Cecilia. *Introduction to Astronomy*. Englewood Cliffs: Prentice-Hall, 1954.

Payne-Gaposchkin, Cecilia, and Katherine Haramundanis. *Introduction to Astronomy*. 2nd edn. Englewood Cliffs: Prentice-Hall, 1970.

Pendleton, Thomas A. "Stephen Greenblatt's Will in the World." *The Shakespeare Newsletter* 54, Nos. 2-3 (2005), 71-2.

Polcaro, V.F., and Martocchia, A. "Supernova astrophysics from Middle Age documents." In *Proceedings IAU Symposium* No. 230. Eds. Evert J.A. Meurs and Guiseppina Fabbiano.

Popper, Karl R. *Conjectures and Refutations: The Growth of Scientific Knowledge*. New York: Harper, 1968.

- - - . *The Logic of Scientific Discovery*. New York: Harper and Row, 1968.

Powell, J.G.F. ed. and trans. *Cicero: Laelius, On Friendship, and The Dream of Scipio*. Warminster: Aris and Phillips, 1990.

Quill, Humphrey. *John Harrison: The man who found longitude*. New York: Humanities Press, 1966.

Quinn, David Beers, ed. *The Roanoke Voyages* 1584-1590. 2 vols. London: Hakluyt Society, 1955.

Rabkin, Norman. *Shakespeare and the Problem of Meaning*. Chicago: University of Chicago Press, 1981.

Radnitzky, Gerard. *Contemporary Schools of Metascience*. Chicago: Henry Regnery, 1973.

Rajan, Tilottama. "'Nothing sooner broke'; Donne's Songs and Sonets as Self-Consuming Artifact." In *John Donne* ed. Andrew Mousley. New York: St. Martin's Press, 1999.

Ramsay, Mary Paton. "Donne's Relation to Philosophy." In *A Garland for John Donne: 1631-1931*. Theodore Spencer, ed. Gloucester, MA: Peter Smith, 1958.

Rebholz, Ronald A. *The Life of Fulke Greville*. Oxford: Clarendon, 1971.

Reynolds, Graham. *English Portrait Miniatures*. Cambridge: Cambridge University Press, 1988.

Richardson, David A. *Dictionary of Literary Biography*. Detroit: Gale Research, *c.*1994, Vol. 136.

Rienitz, Joachim. "Make Glasses to See the Moon Large: An Attempt to Outline the Early History of the Telescope." *Bulletin of the Scientific Instrument Society* No. 37 (1993), 7.

- - - . *Historisch-physikalische Entwicklungslinien optischer Instrumente*. Lengerich: Pabst, 1999.

Ringwood, S.D. Letter. *Bulletin of the Scientific Instrument Society* No. 37 (1993), 9-10.

Roberts, John R. *John Donne: An Annotated Bibliography of Modern Criticism, 1979-1995*. Pittsburgh: Duquesne University Press, 2004.

Rogers, Louis W. *The Ghosts in Shakespeare*. Wheaton: Theosophical Press, 1966.

Ronan, Colin A. "Leonard and Thomas Digges." *Endeavour New Series* 16, No. 2 (1992), 91-4.

- - - . "Postscript concerning Leonard and Thomas Digges and the Invention of the Telescope." *Endeavour New Series* 17, No. 4 (1993), 177-9.

- - - . "There was an Elizabethan Telescope." *Bulletin of the Scientific Instrument Society* No. 37 (1993), 2-3.

Rosen, Edward. *The Naming of the Telescope*. New York: Schuman, 1947.

- - - . "Copernicus published as he perished." *Nature* 241 (1973), 433-4.

Ross, John J. "Shakespeare's Chancre: Did the Bard have Syphilis?" *Clinical Infectious Diseases* 40 (2005), 399-404.

Roston, Murray. *The Soul of Wit: A Study of John Donne*. Oxford: Clarendon Press, 1974.

Roth, Steve. *Hamlet: The Undiscovered Country*. <princehamlet.com>, 2002.

Rowse, Alfred L., ed. *Cymbeline*. By William Shakespeare. Lanham, MD: University Press of America, 1987.

- - - . *Shakespeare The Man*. New York: St. Martin's Press, 1988.

- - - . ed. *The Annotated Shakespeare*. New York: Crown Publishers, 1988, Vol. III.

Roy, Rob. "Shakespeare, No Astronomer." *Event Horizon* 3, No. 2 (1995).

Sandage, Allan, and John Bedke. *The Hubble Atlas of Galaxies*. Washington: The Carnegie Institute, 1994.

Satin, Joseph. *Shakespeare and his Sources*. Boston: Houghton Mifflin, 1966.

Schmidt, Maarten, and Richard F. Green. "Quasar Evolution derived from the Palomar Bright Quasar Survey and Other Complete Quasar Surveys." *The Astrophysical Journal* 269 (1983), 352-374.

Schoenbaum, Samuel. *Shakespeare's Lives*. Oxford: Clarendon Press, 1991.

- - - . *William Shakespeare: A Compact Documentary Life*. New York: New American Library, 1986.

Seeger, Raymond J. *Galileo Galilei: His Life and his Works*. Oxford: Pergamon, 1966.

Seigfried, Hans. "Transcendental Experiments (II): Kant and Heidegger." In *Hermeneutic Phenomenology: Lectures and Essays*. Ed. J.J. Kockelmans. Washington: University Press of America, 1988, pp.123-156.

Seymour, Michael, et al. *Bartholomaeus Anglicus and his Encyclopedia*. Aldershot: Ashgate, 1992.

- - - . *On the Properties of Things: John Trevisa's Translation of* Bartholomaes Anglicus De Proprietatibus Rerum. 2 vols. Oxford: Clarendon, 1975.

Seznec, Jean. *The Survival of the Pagan Gods*. Kingston TN: Pantheon Books, 1953.

Shapley, Harlow. *The Realm of the Nebulae*. New Haven: Yale University Press, 1936.

Shapley, Harlow, and Heber Curtis. "The Scale of the Universe." *Bulletin of the National Research Council* 2, Part 3, No. 11 (1921).

Shirley, John W. *Thomas Harriot: A Biography*. New York: Clarendon Press, 1983.

Showerman, Earl. "Orestes and Hamlet: From Myth to Masterpiece: Part 1." *The Oxfordian* 7 (2004), 89-114.

Simpson, Evelyn M. "Donne's 'Paradoxes and Problems.'" In *A Garland for John Donne*. Ed. Theodore Spenser. Cambridge: Harvard University Press, 1931, pp. 21-49.

Singer, Dorothea W. *Giordano Bruno: His Life and Thought*. New York: Schuman, 1968.

Sluiter, Engel. "The Telescope before Galileo." *Journal for the History of Astronomy* 28 (1997), 223-34.

Smith, G.C. Moore. *Gabriel Harvey's Marginalia*. Stratford-upon-Avon: Shakespeare Head Press, 1913.

Sohmer, Steve. "Certain Speculations on *Hamlet,* the Calendar, and Martin Luther." *Early Modern Literary Studies* 2.1 (1996), 5.1-51.

- - -. *Shakespeare's Mystery Play: The Opening of the Globe 1599*. Manchester: Manchester University Press, 1999.

- - - . "A note on Hamlet's illegitimacy identifying a source of the 'dram of eale' speech (Q2 1.4.17-38)." *Early Modern Literary Studies* 6.3 (2001), 13.1-7.
- - - . "The 'Double Time' Crux in Othello Solved." *English Literary Renaissance* 32, No. 2 (2002), 214-238.
Sokol, Barnett J. "The problem of assessing Thomas Harriot's A briefe and true report of his discoveries in North America." *Annals of Science* 51 (1994), 1-16.
Spenser, Theodore. "Donne and His Age." In *A Garland for John Donne*. Ed. Theodore Spenser. Cambridge: Harvard University Press, 1931, pp. 177-202.
Spiller, Elizabeth. *Science, Reading, and Renaissance Literature, 1580-1670*. Cambridge: Cambridge University Press, 2004.
Spong, John Shelby. *Why Christianity Must Change or Die*. San Francisco: Harper, c.1998.
Stahl, William Harris. *Macrobius: Commentary on the Dream of Scipio*. New York: Columbia University Press, 1952.
Stegmüller, Wolfgang. *Collected Papers on Epistemology, Philosophy of Science and History of Science*. Dordrecht: Reidel, 1977, Vol. II.
- - - . *Rationale Rekonstruktion von Wissenschaft und ihrem Wandel*. Stuttgart: Philipp Reclam, 1979.
Stephenson, Francis Richard, and David A. Green. *Historical Supernovae and their Remnants*. New York: Oxford University Press, 2002.
- - - . "Was the supernova of AD 1054 reported in European history?" *Journal of Astronomical History and Heritage* 6, No. 1 (2003), 46-52.
Stern, Virginia F. *Gabriel Harvey*. Oxford: Clarendon, 1979.
Stock, Brian. *Myth and Science in the Twelfth Century*. Princeton: Princeton University Press, 1972.
Strong, Roy C. *Nicholas Hilliard*. London: Joseph, 1975.
Strong, Roy, C., and V.J. Murrell. *Artists of the Tudor Court: The Portrait Miniature Rediscovered*. London: The Victoria and Albert Museum, 1983.
Strunk, William Jr., and E.B. White. *The Elements of Style*. New York: Macmillan, 1979.
Summers, James Bradley. "Acknowledging Intellectual Debt." *Journal of Nursing Scholarship* 35, No. 4 (2003), 305.
Swank, Lowell James. "Rosencrantz and Guildenstern in London." *The Shakespeare Oxford Newsletter* Vol. 39, No. 2 (2003) 1 and 12-15.
Tey, Josephine. *The Daughter of Time*. New York: MacMillan, 1951.
Thoren, Victor E. *The Lord of Uraniborg*. Cambridge: Cambridge University Press, 1990.
Tilley, Morris Palmer. *A Dictionary of the Proverbs in England in the Sixteenth and Seventeenth Centuries*. Ann Arbor: University of Michigan Press, 1950.
Toomer, G.J. "Ptolemy and his Greek Predecessors." *Astronomy before the Telescope*, ed. Christopher Walker. New York, St. Martin's, 1996, pp. 68-91.
Trimble, Virgina, and Markus Aschwanden. "Astrophysics in 2004." *Publications of the Astronomical Society of the Pacific* 117 (2004), 311-394.

Turner, Gerard L'E. "There was no Elizabethan Telescope." *Bulletin of the Scientific Instrument Society* No. 37 (1993), 3-5.

- - -. "Later Medieval and Renaissance Instruments." In *Astronomy before the Telescope*, ed. Christopher Walker. New York, St. Martin's, 1996, pp. 231-44."

Tuve, Rosemond. *The Zodiake of Life, by Marcellus Palingenius*. Ann Arbor: Edwards Brothers, 1947.

Usher, Peter. "Astronomy and the Canons of Hermeneutics: Mercury and the Elusiveness of Meaning." *The Astronomy Quarterly* 3 (1979/80), 115-124, 171-184.

- - - . "A New Reading of Shakespeare's King John." *Bulletin of the American Astronomical Society* 27 (1995), 1325.

- - - . "Astronomy and Shakespeare's Hamlet." *Bulletin of the American Astronomical Society* 28 (1996), 859.

- - - . "A New Reading of Shakespeare's Hamlet." *Bulletin of the American Astronomical Society* 28 (1996), 1305.

- - - . "Shakespeare's Cosmic World View." *Mercury Magazine* 26, No. 1 (1997), 20-23.

- - - . "Hamlet and the Infinite Universe." *Research Penn State* 18, No. 3 (1997), 6-7.

- - - . "Hamlet's Transformation." *Bulletin of the American Astronomical Society* 29 (1997), 1262.

- - - . "Amleto e l'Universo infinito." *Giornale di Astronomia* 24, No. 3 (1998), 27-30.

- - - . "Harriot, Digges, and the Ghost in Hamlet." *Bulletin of the American Astronomical Society* 30 (1998), 1428.

- - - . "Hamlet's Transformation." *Elizabethan Review* 7, No. 1 (1999), 48-64.

- - - . "Hamlet's Transformation." *A Groat's Worth of Wit* 11, No. 3 (2000), 39-51.

- - - . "Ruin and Revolution in Hamlet." *Bulletin of the American Astronomical Society* 31 (1999), 871.

- - - . "Sixteenth Century Astronomical Telescopy." *Bulletin of the American Astronomical Society* 33 (2001), 1363.

- - - . "New Advances in the Hamlet Cosmic Allegory." *The Oxfordian* 4 (2001), 25-49.

- - - . "Shakespeare's Support for the New Astronomy." *The Oxfordian* 5 (2002), 132-46.

- - - . "Jupiter and Cymbeline." *The Shakespeare Newsletter* 53, No.1, (2003), 7-12.

- - - . "Galileo's Telescopy and Jupiter's Tablet." *Bulletin of the American Astronomical Society* 35 (2003), 1258.

- - - . "Hamlet's Love Letter and the New Philosophy." *The Oxfordian* 8 (2005), 93-109.

van Helden, Albert. *Measuring the Universe*. Chicago: University of Chicago Press, 1985.

- - - . "The Invention of the Telescope." *Transactions of the American Philosophical Society* 67, Part 4 (1977), 1-67.

- - - , trans. *Sidereus Nuncius. By Galileo Galilei*. Chicago: University of Chicago Press, 1989.

Vendler, Helen. *The Art of Shakespeare's Sonnets*. Cambridge, MA: Belknap Press, 1999.

Wagner, Jeffrey K. *Introduction to the Solar System*. Philadelphia: Saunders, 1991.

Wain, John. *Feng*. New York: Viking, 1975.

Warnke, Frank J. *John Donne*. Boston: Twayne, 1987.

Warren, Roger, ed. *Cymbeline*. By William Shakespeare. Oxford: Clarendon, 1998.

Watson, Fred. *Stargazer: The Life and Times of the Telescope*. Da Capo Press, 2004.

Watt, Robert. *Bibliotheca Britannica: A General Index to British and Foreign Literature*. 2 Vols. Constable: Edinburgh, 1824.

Weaver, Warren. *Lady Luck: The Theory of Probability*. New York: Doubleday, 1963.

Webb, Henry J. *Elizabethan Military Science*. Madison: University of Wisconsin Press, 1965.

Westfall, Richard S. *Never at Rest: A Biography of Isaac Newton*. Cambridge: Cambridge University Press, 1983.

Westheimer, Gerald. In *Adler's Physiology of the Eye*. Ed. P.L. Kaufman and Albert Alm. St. Louis: Mosby, 2003, pp. 453-69.

Westman, Robert S. "Three responses to the Copernican Theory." In *The Copernican Achievement*. Ed. Robert S. Westman. Berkeley: University of California Press, 1975, pp. 285-345.

Westra, Haijo Jan. *The Commentary on Martianus Capella's De Nuptiis Philologiae et Mercurii attributed to Bernardus Silvestris*. Toronto: Pontifical Institute of Medieval Studies, 1986.

Wetherbee, Winthrop. *The Cosmographia of Bernardus Silvestris*. New York: Columbia University Press, 1973.

- - - . In *Encyclopedia of Philosophy*. Ed. Edward Craig. London: Routledge, 1998.

Whalen, Richard F. "Cross-examining Leonard Digges on his Stratford Connections." *Shakespeare Oxford Newsletter*. 37, No.1 (2001), 13-15.

Whitney, Charles A. *The Discovery of Our Galaxy*. New York: Knopf, 1971.

Williams, Sheridan. *UK Solar Eclipses from Year 1 to 3000*. Leighton Buzzard: Clock Tower Press, 1996.

Wilson, J. Dover. *What Happens in Hamlet* (1935). Cambridge: Cambridge University Press, 2001.

- - - , ed. *Hamlet*. By William Shakespeare. Cambridge: Cambridge University Press, 1934.

Wimsatt, W.K. Jr., and M.C. Beardsley. "The Intentional Fallacy." *The Sewanee Review* 54, No. 3 (1946), 468-88.

Winnifrith, Alfred. *Men of Kent and Kentish Men*. Folkestone: Parsons, 1913.

Winstanley, Lilian. *Hamlet and the Scottish Succession*. Cambridge: Cambridge University Press, 1921.

Wood, Anthony à, with Philip Bliss. *Athenae Oxoniensis*. London: Rivington et al. 1813.

Wood, Perran. "The Tradition from Medieval to Renaissance." In *The History of Glass*. Eds. Dan Klein and Ward Lloyd. London: Orbis, 1984. pp. 67-92.

Youngson, Robert M. *The Madness of Prince Hamlet and Other Extraordinary States of Mind*. New York: Carroll and Graf, 1999.

Zombeck, Martin V. *Handbook of Space Astronomy and Astrophysics*. Cambridge: Cambridge University Press, 1990.

Zsoldos, E. "On the Origin of the Term RV-Tauri-type." *Observatory* Vol. 113, No. 1117, 305-6, 1993.

Zunder, William. *Elizabethan Marlowe*. Hull: Unity Press, 1994.

Printed in the United States
100998LV00006B/187/A